Invoice/No. 18589
18/3/25

HI £6-35

17

STRATHCLYDE UNIVERSITY LIBRARY
30125 00076616 1

KV-038-730

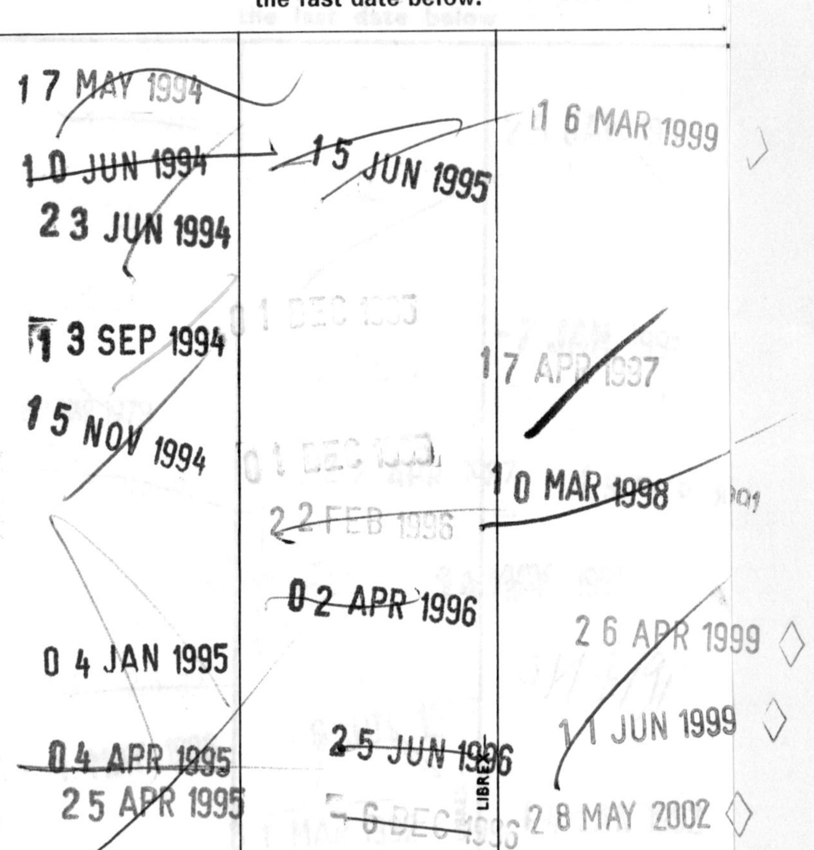

Incineration

Incineration

A State-of-the-Art Study

National Center for
Resource Recovery, Inc.

Lexington Books
D.C. Heath and Company
Lexington, Massachusetts
Toronto London

Library of Congress Cataloging in Publication Data

National Center for Resource Recovery
Incineration: a state-of-the-art study.

Bibliography: p.
1. Incineration. I. Title.
TD796.N36 1974 628'.445 74-6124
ISBN 0-669-94573-0

Copyright © 1974 by National Center for Resource Recovery, Inc.

All rights reserved. No part of this publication may be reproduced or transmitted in any form or by any means, electronic or mechanical, including photocopy, recording, or any information storage or retrieval system, without permission in writing from the publisher.

Published simultaneously in Canada.

Printed in the United States of America.

International Standard Book Number: 0-669-94573-0

Library of Congress Catalog Card Number: 74-6124

628.445
NAT

Contents

ANDERSONIAN LIBRARY
WITHDRAWN FROM LIBRARY STOCK
UNIVERSITY OF STRATHCLYDE

List of Figures

List of Tables

Acknowledgments

The National Center especially acknowledges the efforts of Mr. Morris Zusman, Systems Analyst for the National Center, and Dr. Elmer R. Kaiser, P.E., Consulting Engineer (formerly Senior Research Scientist with New York University). Both men provided the technical expertise that was required to write this volume.

We wish to also express our thanks to the following individuals who reviewed this book and added to its content through their comments: Dr. John H. Fernandes, Coordinator, Environmental Control Systems, Combustion Engineering, Inc.; Mr. Robert S. Rochford, Industrial and Marine Division, Babcock & Wilcox; Mr. Herbert I. Hollander, American Society of Mechanical Engineers; Ms. Monica Maxon, Assistant Information Analyst for the National Center; and Mr. Ronald Vancil, Consulting Economist for the National Center.

Incineration

1 Introduction

Incineration is the process of thermally reducing the volume of solid waste, while producing inoffensive gases and sterilized residue, by the application of the combustion process. In spite of this compact and reassuring definition, however, the potential contribution of incineration to a more satisfactory urban environment is only beginning to be understood. Once hailed as an important advance in sanitary refuse disposal, then castigated as an air and water polluter, incineration is now undergoing intensive technical development to meet new high standards. And while the use of incineration continues to mount, the public at large has tended to mistrust this method of solid waste disposal. On the other hand, current fossil fuel shortages and the recognition of the energy value potential of refuse has spurred interest in advanced systems for incineration.

Incineration has much to recommend it. An effective incinerator can reduce the volume of collected trash by as much as 90 to 95 percent (using high-temperature incineration) while simultaneously changing this vermin- and insect-breeding organic material to an inert ash. As a result, the need for trucking to a remote landfill is not only drastically reduced, but the landfill's capacity is increased nearly twelve times, which also produces significant cost savings.

Still, compared with landfilling or less-desirable open dumping, incineration is a costly procedure requiring a building, mechanical equipment for burning and handling the refuse, a way of minimizing airborne particulate and smoke, ash quenching, storage and removal of the ash, and a working crew to operate the facility on an efficient, economic, and possibly 24-hour basis.

Several of the by-products of incineration can be sold to help offset operating costs. The plant could be a first step in recycling—that is, it would become a point at which easily retrievable and marketable materials could be collected from the waste stream. At the least, scrap metal could be recovered from the residue, or the residue could be used as stable fill material. Another potential result of incineration—heat in the form of steam—has received a great deal of attention and has proven quite successful in European installations.

An incinerator operating at temperatures of 1300 to 2000 degrees F. and at a loading rate that allows time for complete combustion can convert refuse to an inert, sterile, inorganic residue containing metal, glass, and compact ash. With proper cleaning and filtering of the stack exhaust, air pollution would be almost non-existent. And with added attention to the aesthetics of exterior design and landscaping, public approval could generally be assured. If the facility has enough operating capacity to minimize the storage of unburned trash and if the facility is properly ventilated, there would be little of the unacceptable odor or unsightly litter associated with open dumping. Proper scheduling of trucks hauling to and from the incinerator can eliminate community objections to traffic snarls or noise.

Unfortunately, too few municipal incinerators attain all the criteria necessary to meet public approval. More frequently, the plants are operated over-capacity with a backlog of unsightly trash waiting to be burned. Operating under the pressure of a mounting or unpredictable backlog, only partial incineration is achieved, producing a residue containing undesirable unburned organics. Incomplete burning of the combustion gases also contributes to air pollution. In such an ineffective incinerator, the solid residue that must be transported to the landfill area consists of charred trash that is only partially reduced in weight and volume. Thus, the tax-paying citizen who must approve the bond issues to build newer incinerators capable of handling the expanding load is often convinced that the basic process is at fault. This need not be the case—as this study will indicate.

Historical Overview

During the long history of urban communities, it has been common practice to dispose of household, commercial, and industrial solid wastes in dumps. But because of the relatively few dense population centers, modest levels of affluence, and partial salvage or reuse of some wastes, the volume of refuse was small until the last half century. Up until well past the French Revolution, clothes for everyone but the loftiest aristocrats were patched and recycled. Much of society's discards were burned in home stoves and fireplaces or open dumps. Leaves were burned in the open; grass was buried in gardens; table and kitchen waste was fed to the livestock. Smoldering fires at town dumps were also tolerated, but with serious reservations. The open fires at dumps partly reduced the volume of the waste, discouraged rodents and vermin, and extended the life of the dumping space. However, the resultant smoke and odor drifted into inhabited areas and the fires frequently grew out of control.

A major advance in refuse disposal occurred when large incinerators were first developed in England a hundred years ago (1874). Within fifty years, incinerator plants were built in a number of large cities in Europe and America, where they displaced the open dumps with a systematic process for salvaging metal, paper, glass, and ash and then burning the putrescible matter. The generation of steam and electricity from the heat released by the burning waste also came into being on a small scale.

Continuous-feed incinerators are a relatively new innovation. In fact, it was not until 1962 that more continuous-feed incinerators were constructed than batch-feed incinerators. Even with the success of the Atlanta, Georgia, continuous-feed incinerator built in 1941, the idea did not gain strength until the early 1950s, when New York City started building two-section, traveling-grate incinerators. In a later promising development, the first continuous-feed, rocking-grate furnace developed in this country was installed in 1962.

The potential of using waste heat was recognized almost as soon as incineration itself and the first steam-generating refuse incinerator was built in 1898 in New York City. The first water-wall incinerator–boilers were built in Europe during the early 1960s, while American incinerators continued to be built exclusively with refractory-wall furnaces until 1967 when the first U.S. water-wall incinerator was built at the Norfolk Navy Yard.

Combustion

The products of complete or controlled combustion are ideally carbon dioxide and water with small quantities of sulfur dioxide and nitrogen. In practice, some of the nitrogen will be converted to oxides, with nitric oxide being the predominant product. Small amounts of sulfur found in refuse may become sulfur trioxide upon contact of the sulfur dioxide with the atmosphere. Other non-combustibles in the trash will also partially oxidize under the influence of heat. Usually, these will consist of oxides of aluminum and iron, which are not considered hazardous to animal life.

Chemical analysis of trash and the burning process has established the amounts of air (or oxygen) required for complete combustion. Any air fed to the process that is not needed for complete combustion is classified as excess air. If the waste to be reduced is enclosed in a container that receives controlled amounts of heat and oxygen, and the oxygen is insufficient for complete combustion, smoke is produced. Smoke contains liquids and solids in a dispersion of droplets and particles. The fine droplets or aerosols are actually vaporized liquids produced by the heating of combustible material in the trash.

To better understand the process of incinerating, let us begin with a ton (2,000 pounds) of typical municipal refuse and follow it through a modern incineration process. The input to the furnace consists of the 2,000 pounds of refuse, 18,930 pounds of combustion air (assuming 200 percent excess air), and the moisture present in the combustion air (250 pounds); possible input weights are shown in Table 1–1.

The output from the combustion chamber varies among incinerators, but the yields given in Table 1–2 are indicative of good practice—that is, the gases are cooled by a waste heat boiler and cleaned by an electrostatic precipitator. (In the table, these gases are listed in order of their volumetric quantities.)

It will be noted that the four major harmless gases comprise over 99.92

Table 1-1
Typical Input Weights

	Weight Percent	*Lb. per Ton of Refuse*
Refuse		
Moisture	28.0	560
Carbon	25.0	500
Hydrogen	3.3	66
Oxygen	21.1	422
Nitrogen	0.5	10
Sulfur	0.1	2
Chlorine	0.4	8
Metal	7.2	144
Glass, Ceramics, Stones	8.9	178
Ash	5.5	110
Air		
Nitrogen and Inert Gas		14,548
Oxygen		4,382
Moisture		250
Water Vapor from Residue Quench		100
Total		21,280

Source: E.R. Kaiser

Table 1-2
Typical Products of Incineration

	Lb. per Ton of Refuse	*Parts per Million by Volume*
Stack Gases		
Nitrogen and Inert Gases	14,556.5	705,233
Oxygen	3,006.5	128,062
Water Vapor	1,482.8	112,389
Carbon Dioxide	1,738.0	53,542
Carbon Monoxide	5.7	279*
Hydrogen Chloride	6.2	232*
Organic Gases	6.8	123*
Nitric Oxide (NO)	1.7	78*
Sulfur Dioxide	3.0	62*
Total Gases	20,807.2	1,000,000
Solids, Dry Basis		
Residue from Grate	442.8	
Collected Fly Ash, 94% effic.	28.2	
Emitted Fly Ash, 6% Loss	1.8	
Total Solids	472.8	
Total	21,280.0	

*In furnace exit gases, typical values, capable of further reduction.

Source: E.R. Kaiser

percent of the total, while the noxious gases are less than 0.08 percent. Unburned combustibles in the residue have a typical value of 20 to 30 pounds.

In this example, from the incineration of one ton of refuse, the total volume of gas and water vapor discharged varies with the temperature and ranges from 512,760 cubic feet at 500 degrees F. to 780,000 cubic feet at 1,000 degrees F.

To ensure smokeless combustion and to prevent damage to the furnace by molten ash (slag), it is common practice in refractory-wall furnaces to supply about three times as much air (200 percent excess air) as is theoretically required. The excess air dilutes the combustion gases and passes out of the chimney with them. However, some modern incinerators with water-wall furnaces operate with only 40 to 80 percent excess air.

While the conversion of the organic matter to gases is the principal effect of combustion, oxygen, which has united with the metals, increases the weight of the residue.

Heat Recovery

Recovery of the heat value by generating steam is becoming more common—particularly with the advent of water-wall furnaces and the changing composition of the refuse. Complete combustion of the organic matter in the ton of refuse given in the previous example is 9 million BTU (4500 BTU/pound). An additional 0.4 million BTU is usually released by oxidation of half of the metals. This 9.4 million BTU is equal to that released from the burning of 66 gallons of fuel oil.

Also, the heat value per ton is increasing. Plastics, rubber, and wax paper, which are rich in heating units—ranging between 10,000 and 19,500 BTU per pound—are increasing as a percentage of the total refuse. Conversely, food waste and grass clippings, which often contain less than 2,000 BTU per pound because of their moisture content, are on the decrease percentage-wise.[1]

As a result, the steam production in American plants as recently as 1964 ranged as high as 1.8 pounds of steam per pound of refuse but averaged on a continuous basis only about 1.25 pounds per pound. More recently, the Norfolk water-wall incinerator has shown a steaming rate of 3.2 pounds per pound and the Chicago Northwest incinerator a steaming rate of 3.0 during a recent test.

Solid Waste

To fully understand the incinerator, we must also know something of solid waste volume and characteristics. Each of us is a generator of solid waste. We produce, on the average, almost one-half ton, or 1,000 pounds, of such waster per person in a year. The total for our 210 million residents is exceeding 100 million tons per year. This tonnage does not include mining and agricultural wastes, rubble, and excavation rock and soil.

Not all solid waste is discarded in the home. Shopping centers, restaurants, service stations, manufacturing plants, and food processors contribute to our individual shares. But eventually, somehow, our portion of this refuse must be burned, buried, or recycled.

One major constitutent of our national solid waste is paper and paper products. Newsprint, cardboard, and other paper account for about 35 to 45 percent of the weight of all municipal solid waste. Metals account for another 9 percent. Food or garbage content of solid waste is approximately 14 percent, and glass is another 10 percent. Other constitutents of solid waste are yard refuse, textiles, wood, rubber, and plastics.

The percentages of the materials in solid waste will change with the season, socio-economic makeup of the neighborhood, geographical location, and other variables. For example, foliage from palm trees may be a problem during certain months in Southern California. Cans and bottles add to the load of waste produced in certain neighborhoods during hot months. Christmas trees and wrapping paper cause an increase during late December and early January.

Seasonal variations of yard waste are an important factor because of the high moisture content of yard waste. In fact, moisture content, influenced mainly by rain and air humidity, is the major day-to-day variable in refuse analysis. The moisture content can vary widely among the individual refuse components that arrive at the incinerator; for example, paper may contain 22 percent moisture and grass, 46 percent.

The National Center for Resource Recovery recently completed an investigation of the average content of municipal solid waste. The results of this study are shown in Table 1-3.

Over one-half of the municipal refuse is truly combustible; the other fraction is made up of inerts and moisture that merely evaporates during incineration.

Since at least one of the objectives of incineration is volume reduction, the space occupied by mixed refuse under various conditions must also be considered. Loose refuse in waste baskets and bins weighs only about 150 pounds per cubic yard. Compactor-type collection trucks compress that to about 450 pounds per cubic yard. When the trucks discharge the refuse at a delivery point—such as a landfill or storage pit of a municipal incinerator—the refuse has a bulk density of 250 to 300 pounds per cubic yard. When this in turn becomes the unburned residue of the incineration process, that density has increased to 1,000 pounds per cubic yard.

Another characteristic of municipal refuse is its range of particle sizes. Usually, refuse collected for incineration is limited to lengths of pieces of

Table 1-3
NCRR Conclusion of Composition of the Municipal Solid Waste Stream

	Municipal Solid Waste Stream, Excluding Rural, as Received (by Wet Weight)
	% of Total
Paper	43.0
Glass	10.0
Ferrous	8.0
Aluminum	0.7
Other Non-ferrous	0.3
Textiles	2.0
Rubber	1.0
Plastics	2.0
Other	33.0
Total	100.00

three feet, and thicknesses of wood of three inches. The length restrictions are to prevent clogged refuse chutes, while the limit on wood thickness is to permit complete combustion in the 30 to 45 minutes usually available in the incinerator furnace. However, this rule is not always strictly enforced.

The Incinerator Institute of America has developed a number of standards on incinerator design and practice. Among them is a widely used classification of refuse, particularly for incinerators designed to burn less than one ton per hour. The Institute also provides a useful tabulation of refuse densities and the quantities of waste typically generated at inhabited buildings. These tabulations are shown in Tables 1-4, 1-5, and 1-6.

Intelligent manipulation of these data with historical records can result in a reasonable calculation of a community's incineration needs. Population growth and increases in industrial output of various materials that will eventually become solid waste can be added to the other data to give a projected outlook for long-range planning.

Naturally, any steps to salvage several or most of the components of municipal waste would reduce the quantity and chemical composition of the waste to be incinerated. For example, the removal of the corrugated boxes and newspaper from the waste stream for recycling to paper mills would not only reduce the combustible matter to be burned but also automatically increase the proportion of moisture content and non-combustibles in the remaining refuse. Salvaging glass and metal *before* incineration would increase the proportion of combustibles and moisture in the remaining waste.

The recovery of metals and glass from the residue *after* incineration has been investigated by the Bureau of Mines on a pilot-plant scale at College

Table 1–4
Classification of Wastes To Be Incinerated

Classification of Wastes								
Type	*Description*	*Principal Components*	*Approximate Composition (% by Weight)*	*Moisture Content (Percent)*	*Incombustible Solids (Percent)*	*BTU Value/lb. of Refuse as Fired*	*BTU of Aux. Fuel per lb. of Waste to be included in Combustion Calculations*	*Recommended Min. BTU/hr. Burner Input (per lb. Waste)*
0*	Trash	Highly Combustible waste, paper, wood, cardboard cartons, including up to 10% treated papers, plastic or rubber scrap; commercial & industrial sources	Trash–100%	10%	5%	8500	0	0
1*	Rubbish	Combustible waste, paper, cartons, rags, wood scraps, combustible floor sweepings; domestic, commercial and industrial sources	Rubbish–80% Garbage–20%	25%	10%	6500	0	0
2*	Refuse	Rubbish and garbage; residential sources	Rubbish–50% Garbage–50%	50%	7%	4300	0	1500
3*	Garbage	Animal and vegetable wastes, restaurants, hotels, markets; institutional, commercial, & club sources	Garbage–65% Rubbish–35%	70%	5%	2500	1500	3000

4	Animal Solids and Organic Wastes	Carcasses, organs, solid organic wastes; hospital, laboratory, abatoirs, animal pounds, & similar sources	100% Animal & Human Tissue	85%	5%	1000	3000	8000 (5000 Primary) (3000 Secondary)
5	Gasesous, Liquid or Semi-Liquid Wastes	Industrial process wastes	Variable	Dependent on predominant components	Variable according to wastes survey	Variable according to wastes survey	Variable according to wastes survey	Variable according to wastes survey
6	Semi-Solid and Solid Wastes	Combustibles requiring hearth, retort, or grate burning equipment	Variable	Dependent on predominant components	Variable according to wastes survey	Variable according to wastes survey	Variable according to wastes survey	Variable according to wastes survey

*The above figures on moisture content, ash, and BTU as fired have been determined by analysis of many samples. They are recommended for use in computing heat release, burning rate, velocity, and other details of incinerator designs. Any design based on these calculations can accommodate minor variations.

Source: Incinerator Institute of America, *I.I.A. Incinerator Standards,* p. 5A.

Table 1–5
Refuse Densities (Average Weight of Refuse)

Type	*Pounds per Cubic Foot*
Type 0 Waste	8 to 10
Type 1 Waste	8 to 10
Type 2 Waste	15 to 20
Type 3 Waste	30 to 35
Type 4 Waste	45 to 55
Garbage (70% H_2O)	40 to 45
Magazines & Packaged Paper	35 to 50
Loose Paper	5 to 7
Scrap Wood and Sawdust	12 to 15
Wood Shavings	6 to 8
Wood Sawdust	10 to 12

Source: Incinerator Institute of America, *I.I.A. Incinerator Standards,* p. 4A.

Park, Maryland. Based on their encouraging results, application of this process is planned for the City of Lowell, Massachusetts, with the assistance of a $2.4 million EPA grant.

Similarly, some incinerators can burn the so-called "filter cake" that results from sewage treatment plants, which remove solids from sewage before discharging the water to streams, lakes, or oceans. These solids are recovered as a "filter cake" containing 75 to 80 percent moisture (20 to 25 percent solids). A typical analysis of dried sludge cake from municipal sewage is shown in Table 1–7. Incinerators that employ multiple hearths or fluidized sand beds can burn this sludge cake on a continuous basis with the assistance of some auxiliary oil or gaseous fuel.

All communities produce some solid waste that is burnable but too large to be acceptable in incinerators built for normal day-to-day refuse. Included in this Oversize Bulky Waste (OBW) category are tree branches, tree trunks and stumps, old mattresses and furniture, demolition lumber, truck tires, boxes, skids, and pallets. Such wastes may constitute 5 or more percent of the total solid waste of the community, particularly where an urban renewal program is underway or where new developments are underway in wooded areas. Open burning of such materials produces smoke and flying ash over a wide area. Burial is unsatisfactory because the waste does not compact sufficiently to prevent underground fires.

The combustible fraction of OBW consists mainly of wood, paint, plastics, textiles, and rubber. The residue of ash from a burned log is only about 2 percent of the log, while 80 percent of the weight of metal furniture remains after the plastic surfaces and upholstering have been burned off. Table 1–8 is a representative analysis of mixed OBW from municipal collections. Such waste may be shredded and burned with regular municipal refuse in conventional incinerators, or it may be burned in special OBW furnaces without shredding.

Table 1-6
Waste Generated at Various Types of Buildings

Classification	*Building Types*	*Quantities of Waste Produced*
Industrial Buildings	Factories	Survey must be made
	Warehouses	2 lbs. per 100 sq. ft. per day
Commercial Buildings	Office Buildings	1 lb. per 100 sq. ft. per day
	Department Stores	4 lbs. per 100 sq. ft. per day
	Shopping Centers	Study of plans or survey required
	Supermarkets	9 lbs. per 100 sq. ft. per day
	Restaurants	2 lbs. per meal per day
	Drug Stores	5 lbs. per 100 sq. ft. per day
	Banks	Study of plans or survey required
Residential	Private Homes	5 lbs. basic & 1 lb. per bedroom
	Apartment Buildings	4 lbs. per sleeping room per day
Schools	Grade Schools	10 lbs. per room & ½ lb. per pupil per day
	High Schools	8 lbs. per room & ½ lb. per pupil per day
	Universities	Survey required
Institutions	Hospitals	15 lbs. per bed per day
	Nurses or Interns Homes	3 lbs. per person per day
	Homes for Aged	3 lbs. per person per day
	Rest Homes	3 lbs. per person per day
Hotels, etc.	Hotels-1st Class	3 lbs. per room & 2 lbs. per meal per day
	Hotels-Medium Class	1½ lbs. per room & 1 lb. per meal per day
	Motels	2 lbs. per room per day
	Trailer Camps	6 to 10 lbs. per trailer per day
Miscellaneous	Veterinary Hospitals Industrial Plants Municipalities	Study of plans or survey required

Source: Incinerator Institute of America, *I.I.A. Incinerator Standards,* p. 4A.

Standard Practice

In the face of all this welter of solid waste, many cities are handling the problem in a way that can be simplified to the flow chart shown in Figure 1-1. It indicates the generally accepted practice of removal and disposal of the by-products of an affluent society. In this chart, it should be noted that incineration of solid waste is not so much an alternative to landfill as it is a means of changing the state of the disposed material as well as decreasing its volume and

Table 1-7
Typical Analysis of Dried Sludge

	Weight Percent Dry Basis
Carbon	34.74
Hydrogen	5.94
Oxygen	27.86
Nitrogen	1.04
Sulfur	0.31
Chlorine	0.40
Non-combustibles (Ash)	29.71
Total	100.00

Calorific Value: 6,255 BTU per pound

Source: E.R. Kaiser

Table 1-8
Typical Analysis of Oversized Burnable Waste

Moisture	15.0%
Carbon	38.0
Hydrogen	5.0
Oxygen	28.0
Nitrogen	0.4
Sulfur	0.2
Chlorine	0.4
Metal, Glass, Ash	13.0
Total	100.0%

Calorific value of organic matter: 7,500 BTU per pound

Source: E.R. Kaiser, D. Kasner, and C. Zimmer, "Incineration of Bulky Refuse III," *Proceedings of the National Incinerator Conference* (1972), ASME, pp. 338–346.

weight. Thus, the process can be considered as a pre-conditioner before final disposition.

Bibliographic Note

Of all that has been written on the subject of incineration, there are a handful of reports that the serious reader should turn to first. Richard C. Corey's *Principles and Practices of Incineration* (New York: Wiley, 1969) is a collection of ten chapters by various experts that cover the subject in precise mathematical and scientific detail. Although a valuable guide to the subject, it seems to have been prepared for designers rather than municipal planners. The book deals with municipal solid waste, as well as industrial and radioactive wastes.

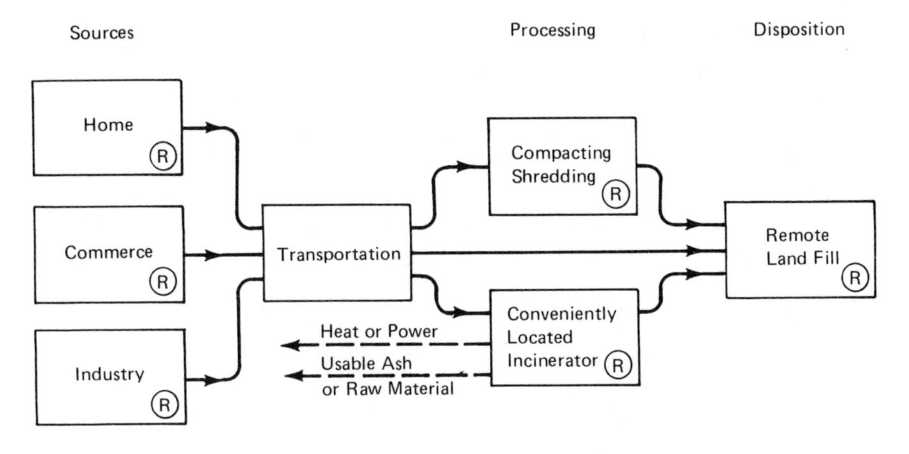

Figure 1-1. From Collected Solid Waste to Landfill. Source: Midwest Research Institute.

Of more universal appeal is the short (64 pages) *Handbook on Incineration: A Guide to Theory, Design, Operation, and Maintenance,* by Frank L. Cross, Jr. (Westport, Conn.: Technomic, 1972). Its economic use of text and artwork make a clear and efficient introduction to the subject.

Similarly, Junius W. Stephenson's "Incineration–Past, Present, and Future," presented at the 1969 ASME Winter Annual Meeting, introduces the topic from a historical perspective. On the basis of an historical and technological survey, it discusses possible future developments.

An Arthur D. Little Report, *Systems Study of Air Pollution from Municipal Incineration,* in three volumes by Walter R. Niessen *et al.* (1970), is also an important study of incineration, although it more properly deals with air pollution problems and research needs.

The Treatment and Management of Urban Solid Waste, edited by David Gordon Wilson (Westport, Conn.: Technomic, 1972), contains a thorough and dependable chapter on "Municipal Incineration." It presents a great deal of concise, practical information.

Last, *Incinerator Guidelines–1969,* written by Jack DeMarco *et al.* and published by the U.S. Public Health Service (Publication No. 2012), is an invaluable review of the important facts entailed by a planned incinerator installation.

In addition to these larger-format publications, a great many shorter papers are presented in various issues of *Proceedings of the National Incinerator Conference* (American Society of Mechanical Engineers, 345 East 47th Street,

N.Y., N.Y., 10017). The conferences have been held every two years since 1964 and volumes for 1970 and 1972 are still available.

Combustion Engineering Corp., Windsor, Connecticut, publishes an annotated bibliography of writings on incineration. The bibliography originally appeared in 1963 and has been updated three times, the most recent supplement being that for 1970.

The Bibliography at the end of this volume presents a somewhat more selected list for more detailed reading on incineration.

Aim of the Study

This volume is intended as an introduction to the concept and practices of incineration for the general reader—and hopefully for the many municipal officials who are thinking of incineration as a solution to their solid waste problem. As such, it surveys the literature and state-of-the-art and refers the reader to other articles and books for more detailed, technical, or specialized knowledge. Significant portions of this study were drawn from the previously mentioned documents.

2 Current Incineration Practice

Centrally located *city* plants are the focus of most of today's refuse incineration. Although the concept finds its greatest acceptance in fairly large cities, groups of smaller communities may also join in building and operating an incinerator plant. Few plants are built at present with daily throughputs of less than 100 tons. Incinerators with two-, three-, and four-furnace units and a total capacity of 500 to 1,200 tons per day are typical.

Incinerators are usually custom-designed, under contract with the municipality, by consulting engineering firms which specialize in this field. They are built by private contractors, with components such as the stoker furnace and boiler builders supplied by a few manufacturers.

At present, there are some 300 municipal incinerators being used in 190 cities in the United States. But many other incinerators have been abandoned in recent years, either because of increased cost of operation and maintenance, or because the anticipated capital expenditures required to make them conform to newer air pollution standards are not economically justifiable. At the same time, new incinerators, which incorporate new technology and provide increased efficiency and pollution control, are being designed and constructed in the United States and abroad, and some older incinerators are being modernized to meet new air pollution standards.

Because of the incinerator's high cost, often a decision to build one is reached only as the result of depletion of landfills or increasing distances of landfill sites from the area where the refuse is generated. Rodney R. Fleming, in the February 1966 issue of *The American City,* cited the example of Delaware County, Pennsylvania: "Although only 185 square miles in area, this urban county embraces 49 municipalities ranging in population from 500 to 90,020. Landfill requirements would have totaled 1,000 acres over the next 20 years and these sites were rapidly disappearing, so county officials constructed three 500-ton-per-day incinerators to dispose of the refuse at a total cost of $6 million, about $4,000 per ton of rated capacity." [1]

Fleming makes further note of the disappearance of conveniently located landfill sites and cover material; he feels this to be the reason for the construction of more new incinerators. This, of course, is especially true in the environs of the large metropolitan areas where available land brings high prices. Low-priced land in these areas is usually so remote that the transportation costs are becoming prohibitive. It is possible that the air pollution created by trucks hauling many loads of refuse to a remote landfill could compare with that created by the incinerator.

Planning for an Incinerator

Municipal incineration systems require planning of a complexity that equals or surpasses that given to public parks, transportation, education, and law enforcement. Legislators are usually more familiar with these latter functions of a municipality because there is a historical basis to provide some guidelines and point out the pitfalls. But in many areas, the concept of cooperative municipal incineration and trash-handling is so new that city governments need help from state and federal officials.

The large capital investment implied by incineration is of like concern to most cities. And what is more important, most planners are aware that there is a real possibility of making a wrong decision about any of the many components in an incineration system. They know too that this is the kind of mistake that cannot be remedied as easily as changing one-way streets, installing another pump, or adding teachers and classrooms.

The case for constructing a large incinerator rather than numerous small municipal units is well supported. For example, the operational crews of many smaller plants cause a duplication of effort that increases the overall system costs. With automated procedures, one trained crew per shift can better control a single plant than numerous partially trained crews can control individual smaller plants. If power is to be generated as a by-product, then a substantial amount of material is required and a large plant is necessary.

There are other advantages to disposal districts. Supporting funds from federal and state agencies are more likely to be available to a cooperative effort than to many splinter operations. Public-owned utilities and sewage disposal districts have proven the feasibility and practicality of such systems and have provided models of financial support methods as well as means of gaining citizen approval.

Metropolitan disposal districts can also provide a cohesiveness of control not available from fragmented approaches. All the refuse is treated in a similar manner. That is to say, one suburban community is not burning trash in a sub-standard system and exhausting solids into the atmosphere where they are airborne to another community with a properly operating system. Also, in this way, future planning and forecasting of changes in the trash load can be made on an organized basis.

Another advantage of a metropolitan disposal district is site selection. While it is doubtful that every community within the district will agree on a single site, it is easier to reach agreement on one site than it is on many sites.

F.L. Heaney lends support to the argument for a regional incineration district and states that incineration efficiency and consequently economic viability are possible with such systems.[2] He further notes that less than half a dozen metropolitan areas in the United States have established any kind of regional refuse disposal system. To make a comparison, he considers a municipality of 40,000 persons that operates its own system and contrasts this with the situation in which a city joins in with neighboring cities to form a district of 160,000 persons. The analysis indicates that a savings of about 40 percent can be achieved by using the regional concept. Even greater savings would be possible in the case of larger districts with higher population densities.

The availability and cost of capital to a municipality is an important consideration when incineration is being investigated. Such plants must be paid for by some planned funds; bonds are often the answer. Unless commerce or industry is required to pay for the services on a per-ton basis, the only other means of operation are through taxes, individual fees, or some recovery process. Increasing attention is being paid to the utilization of waste heat for production of steam, for direct use by municipalities, industry, and electric utilities as a source of income.

Site Location and Community Acceptance

The factors relating to the selection of a site for a municipal incinerator may be divided into two relevant groups. One is suitability. This covers such important points as access, proximity to a landfill, services (gas, water, electricity), room for expansion, foundation conditions, and drainage. The second group of factors is somewhat more vague and is concerned with the acceptance—by the public—of the presence and operation of the plant. In this, the importance of topography cannot be ignored. No matter how effective the incinerator, some gases and particulates will be dispersed into the air. Locating an incinerator so that its exhaust disperses over the open country, or water, is preferable to locating it so that exhaust disperses over a residential area.

The traffic pattern created by vehicles going to and from a municipal incinerator should be of like concern. Close coordination with the city managers of streets, fire and police departments, public (or private) transit companies, and other users of the street and freeway systems is important. An idealized physical location would require the least time and distance for all collection crews. Public acceptance will also be eased if the morning and evening commuters do not have to compete with refuse trucks. Special access roads and waiting areas may be required and will add to the capital cost of the system. The use of transfer stations and compactor trucks can reduce the flow of trucks into a facility.

The following suggestions on selecting a site are taken from government guidelines:

1. Choose a site where construction can conform with existing and planned neighborhood character. In general, industrial and commercial areas are more compatible with incinerators than residential areas. An incinerator plant is usually classed as heavy industry and the evaluation of its location should reflect this. Too frequently the vacant land surrounding an incinerator is later developed for residential or other restricted use, which creates conflict. To avoid potential conflict, the undeveloped surrounding land should be zoned for industrial use.
2. Avoid choosing a site that may conflict with other public buildings. The noise, lights, and 24-hour work-day of normal incinerator operation preclude locating it near a hospital, and heavy truck traffic makes incinerator location near schools undesirable. Centralized public works operations are desirable. Often an incinerator plant can be advantageously located near a sewage treatment plant so that technical services may be shared. There may be economies in locating the incinerator near a garage where vehicle repair facilities and personnel can be shared.
3. Where conflict with neighborhood character is unavoidable, the screening effects of a wall or planting can reduce adverse effects and gain public acceptance. Good architectural design is itself a major asset in overcoming potential neighborhood objection.
4. Institute an effective public relations program. Before full site and design decisions are made, proposals and plans should be presented through the press and for discussion at public meetings. This would serve to demonstrate management response to community desires and a capability for operating an acceptable facility. Presentation of alternatives along with rationale for incineration may be supported by graphic examples and site visits to successfully operating facilities.[3]

In addition to careful selection of the location for an incinerator plant, attention should also be given to other factors in planning the overall concept of the operation. Since the plant will become obsolete after a number of years, consideration should be given to the use to be made of the site. The development of the area surrounding an incinerator as a recreational park has often been suggested. This plan should cover a 20- to 30-year period, during which landfill areas in the vicinity of the plant would be filled, covered, and converted into a variety of recreational areas.

3 Current Incinerator Design

Every municipal incinerator contains certain basic elements—feed system, combustion system or chambers, exhaust system, a means of residue removal, and apparatus for control of air pollution. Heat or power recovery systems are also receiving consideration in large municipal incinerators. There are, of course, other peripheral elements to the incinerator. Shredders, compactors, collection vehicles, pre- or post-incineration sorters for recovery of products for recycling, residue-removal vehicles, landfill apparatus, and land itself are other considerations in system planning.

The method of feeding refuse to an incinerator is commonly used to place the system into one of two very general categories. In a *continuous-feed* system, a steady flow of refuse is achieved. In *batch-feeding,* large quantities of refuse are periodically placed into the furnace from a storage pit by cranes with buckets or grapples. Batch-fed systems are further subdivided and most of these fall into the *circular* type, where rabble arms spread the refuse over a circular grate, or *rectangular* type, usually with a rocking grate.

Most of the new, large incinerators are designed to feed waste into the furnaces continuously. Three types of grates, shown in Figure 3-1, are common in continuous feed furnaces—reciprocating grates, rocking grates, and traveling grates. A fourth type of grate, referred to as the Dussledorf drum grate, is being used in Europe, but as yet has not been used in the United States. Note that the Dusseldorf grate, shown in Figure 3-2, is being used with a waste heat boiler.

Figures 3-3 and 3-4 illustrate circular-grate and rectangular-grate furnaces for batch-fed rather than continuous-fed systems. (An important point to note with each of these is that the charging opening on each is mechanically sealed to prevent air from entering or escaping from the chamber.)

Figure 3-5 illustrates how the grates typically fit in a continuous-feed incinerator. The figure is a sectional view of the East New Orleans incinerator, which uses a reciprocating grate, but for purposes of this discussion, any of the aforementioned grates installed in an incinerator would schematically look alike.

In the East New Orleans illustration, Figure 3-5, there are several noteworthy components. The charging chute (3) is sealed to keep air out by piling the refuse to the top of the chute by the overhead crane (2). Refuse moves along the grates (9) into the combustion chamber (4), where the volatile gases are driven off and burned completely in the secondary combustion chamber (5).

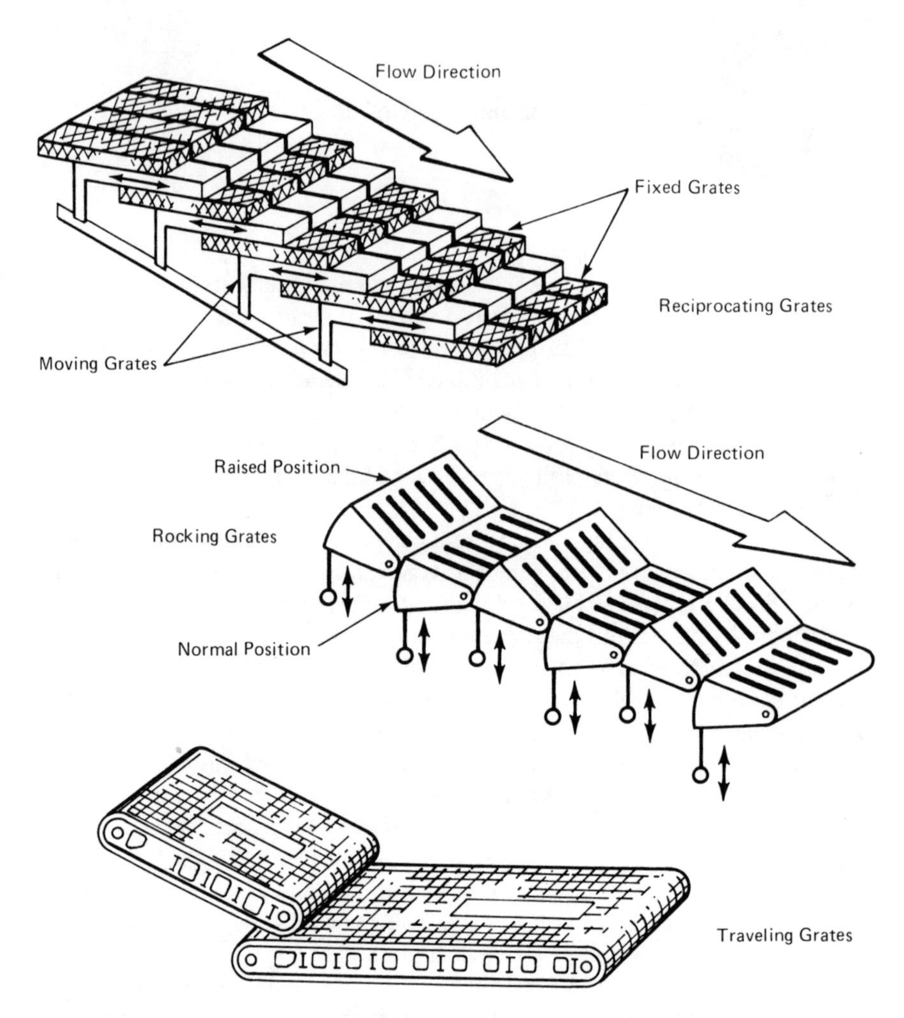

Figure 3-1. Types of Grates for Continuous-Feed Furnaces. Source: DeMarco *et al., Incinerator Guidelines–1969*, pp. 30-31.

Non-volatile matter burns in the primary combustion chamber, while the residue falls into the ash hopper (10). For wet refuse, combustion is supported by the natural gas burners (6). Underfire air (7) is forced up through the burning pile; overfire air (8) ensures high turbulence and complete combustion. In the water spray chamber (11), the hot products of combustion are cooled to where air pollution control equipment can be used. Following the spray chamber, where the gases are cooled to about 600 degrees F., the gases enter the subsidence chamber (13), where the larger particulate settles out. The gas then goes

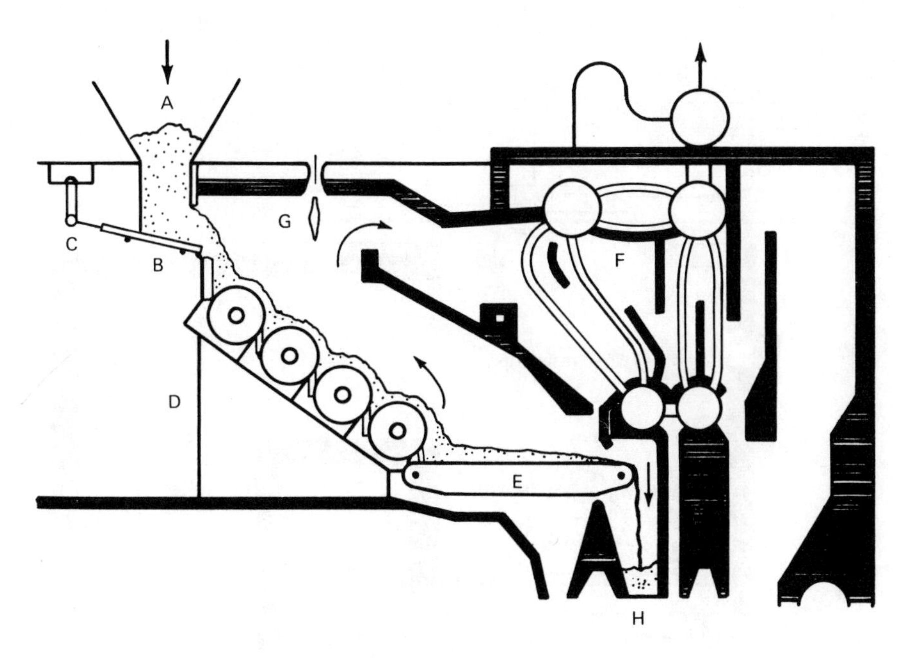

Figure 3-2. The Dusseldorf Incinerator. Source: Stabenow, "European Practices in Refuse Burning," *Proceedings of the National Incinerator Conference* (1964), ASME, p. 107.

to the dust collector (15) and out the stack (17). The overall gas flow patterns are maintained by the induced draft fan (16). Note that there is no waste heat recovery.

Figure 3-6 illustrates a plant being built in Nashville for operation in mid-1974. This plant will have a waste heat boiler with no need for a spray chamber, since the boiler will cool the combustion gases and at the same time generate steam for both heating and cooling.

Another variation is the rotary kiln furnace, shown in Figure 3-7. This furnace is similar to other refractory continuous-feed furnaces, except that it uses a rotary kiln for final burnout in addition to its grates. That kiln, which is sloped to prevent waste buildup, also rotates slowly during the burning process to agitate the unburned refuse.

Obviously, a great many combinations of grate types and furnace types are available for incinerators. A valuable summary of this topic is to be found in "Furnace Configuration" by Frank L. Heaney, in the *Proceedings of the Nation-*

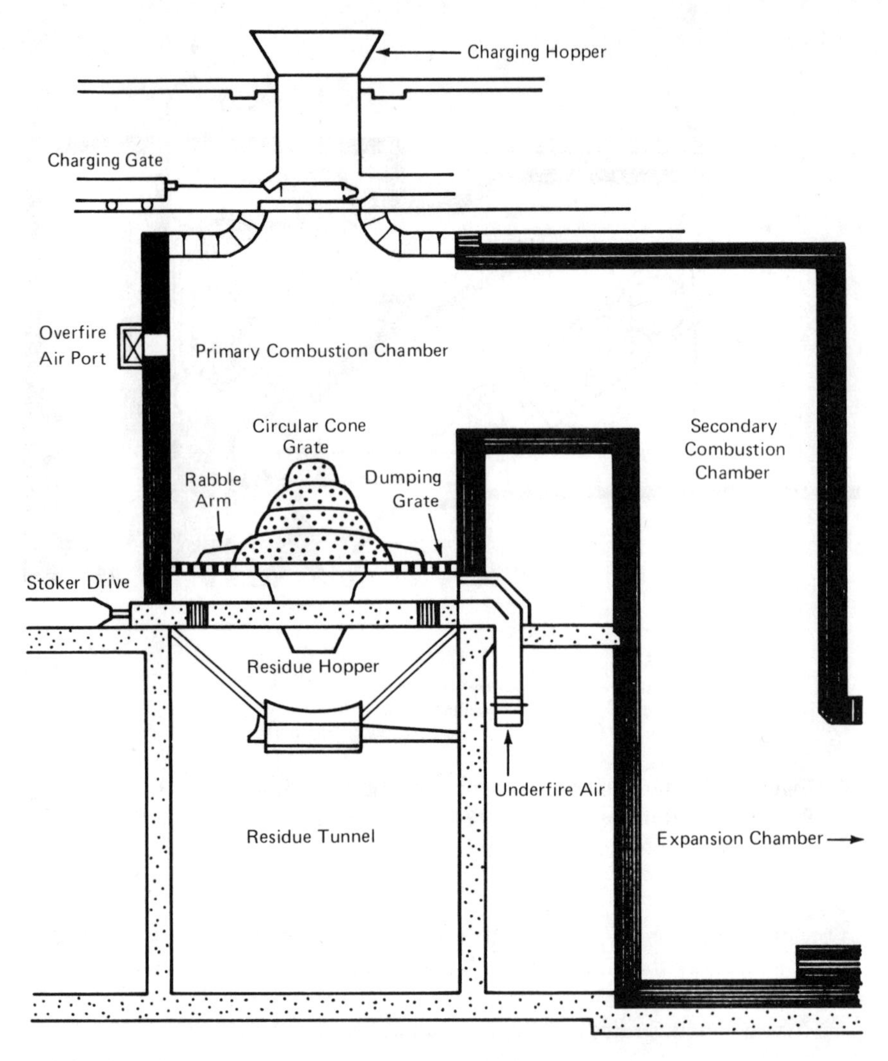

Figure 3-3. Circular Furnace. Source: DeMarco *et al., Incinerator Guidelines–1969*, p. 27.

al Incinerator Conference (1964) which discusses the most prevalent combinations of incinerator elements. The article presents the construction and operation and shows cut-aways of cylindrical batch-fed furnaces, rectangular batch-fed furnaces, continuous-feed traveling grates, continuous-feed rotary kilns, continuous-feed rocking grates, and continuous-feed reciprocating grates.

At this point, it is apparent that incinerator design has become a fairly sophisticated discipline—with a nomenclature and jargon all its own. In this

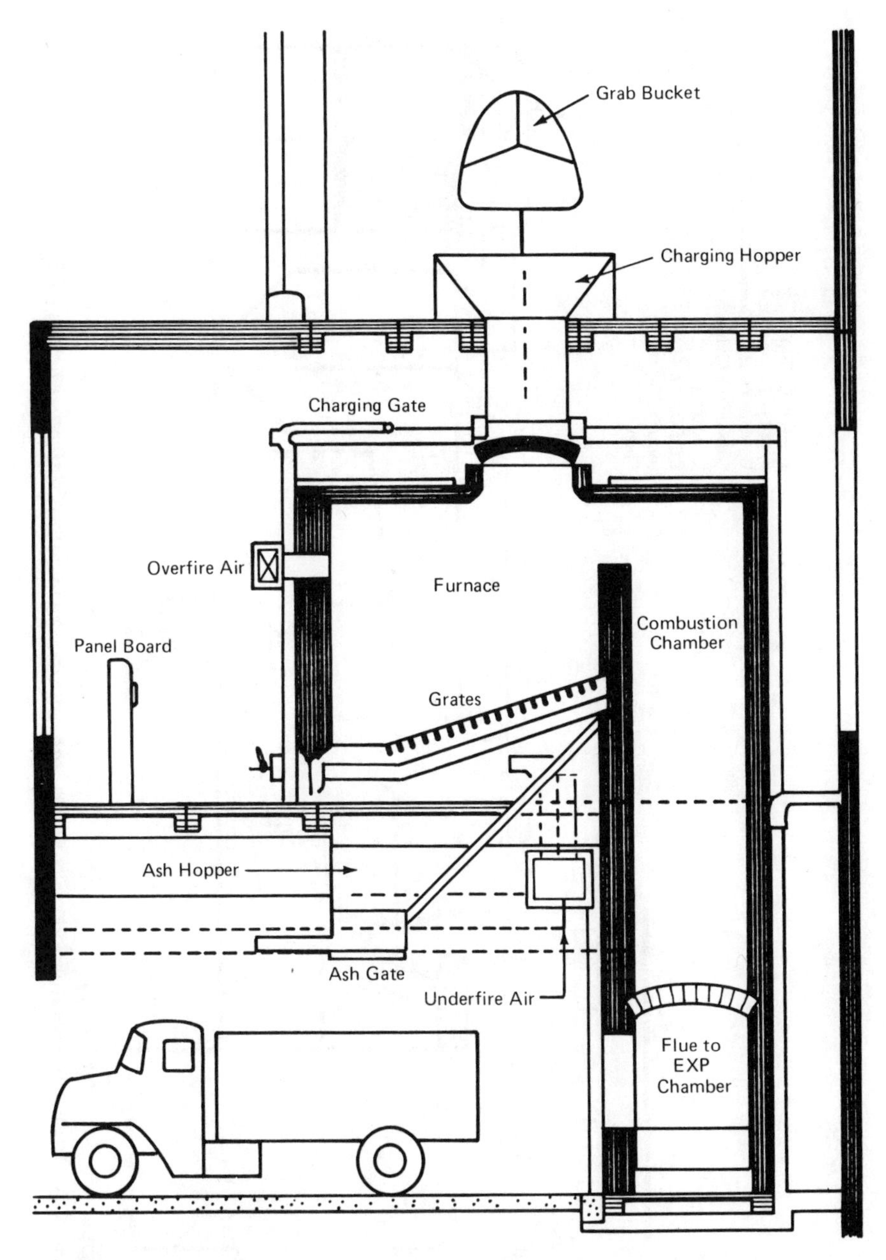

Figure 3-4. Rectangular Furnace. Source: F.I. Heaney, "Furnace Configuration," *Proceedings of the National Incinerator Conference* (1964), ASME, p. 53.

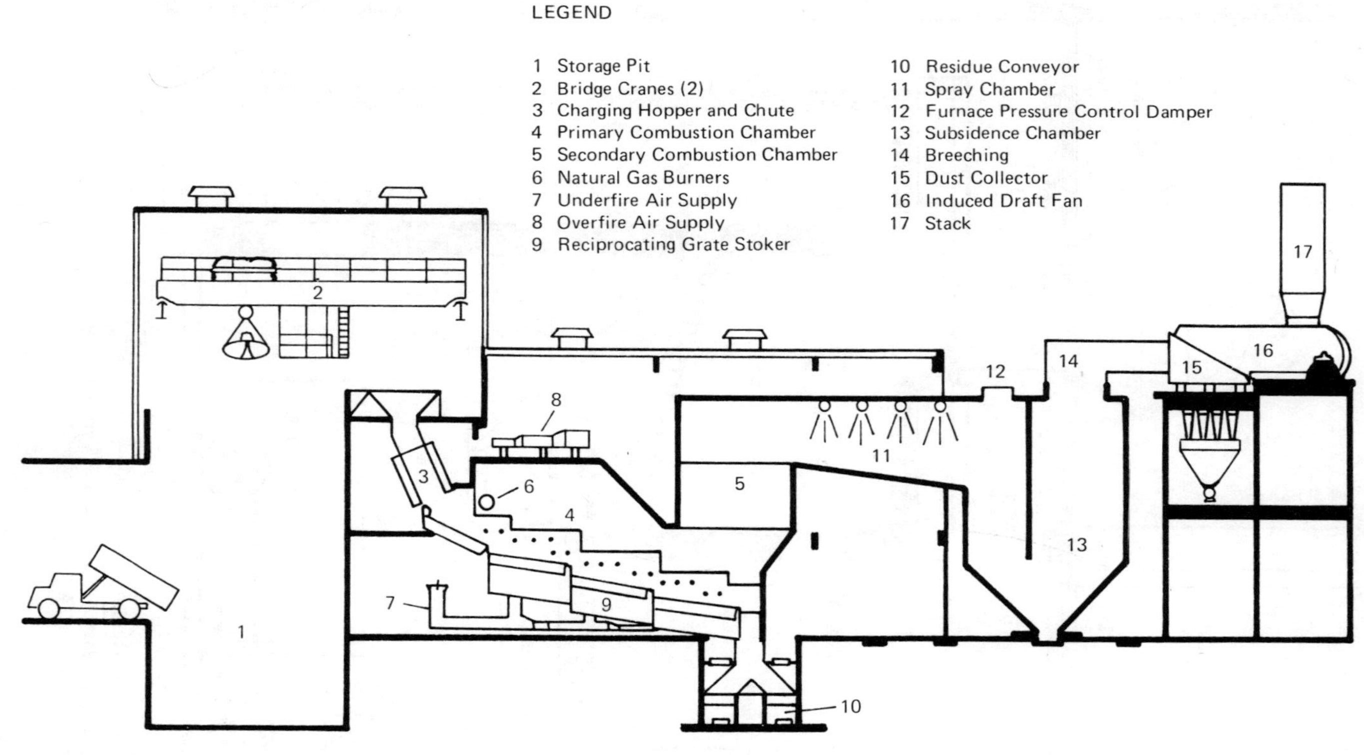

Figure 3-5. East New Orleans Incinerator. Source: T.C. Heil, "Planning, Construction, and Operation of the East New Orleans Incinerator," *Proceedings of the National Incinerator Conference* (1970), ASME, p. 143.

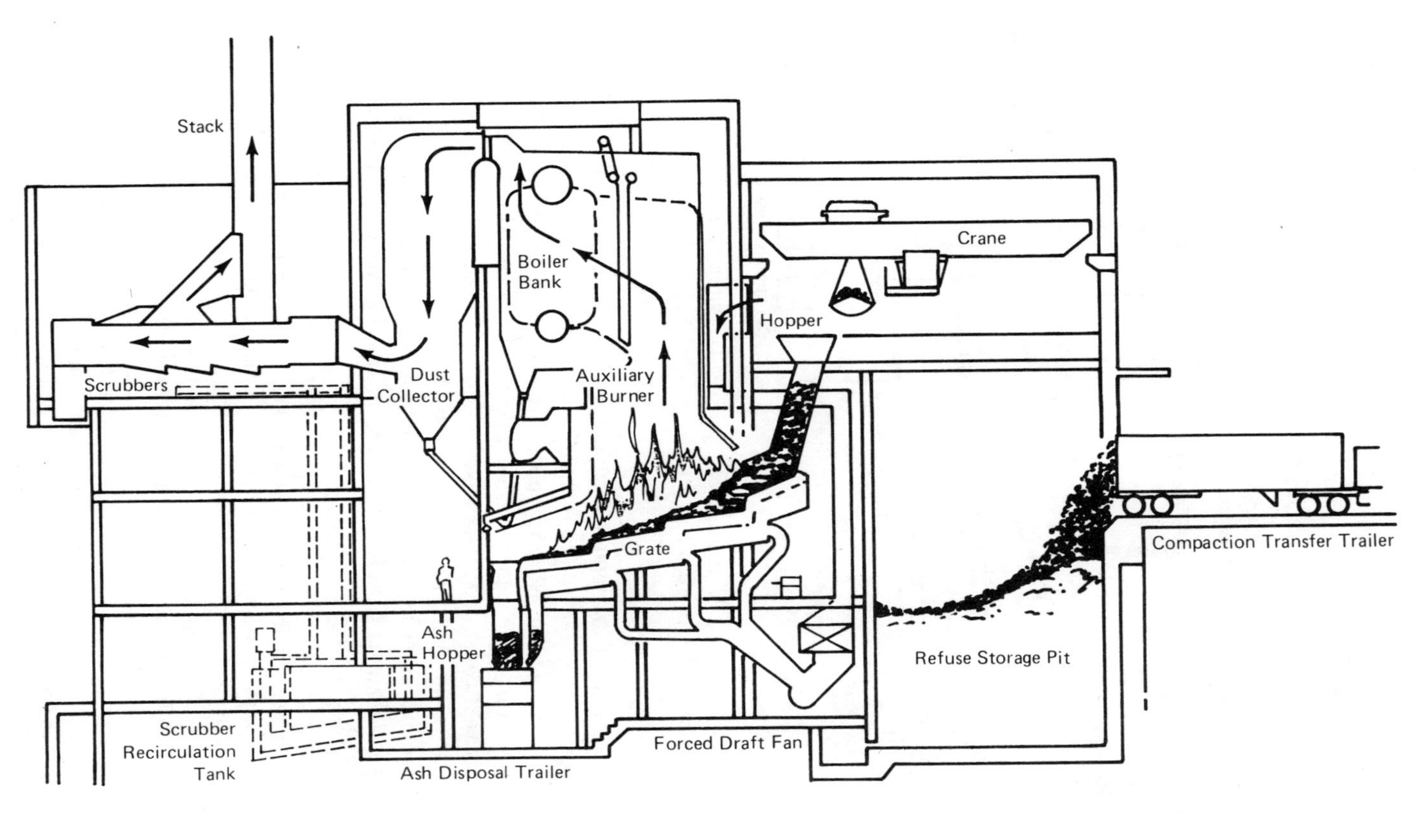

Figure 3-6. Nashville Incinerator–Boiler. Source: "Using Wastes as Fuel, Coty Plant Will Provide Heating and Air Conditioning," *Solid Wastes Management Magazine*, September 1972, p. 23.

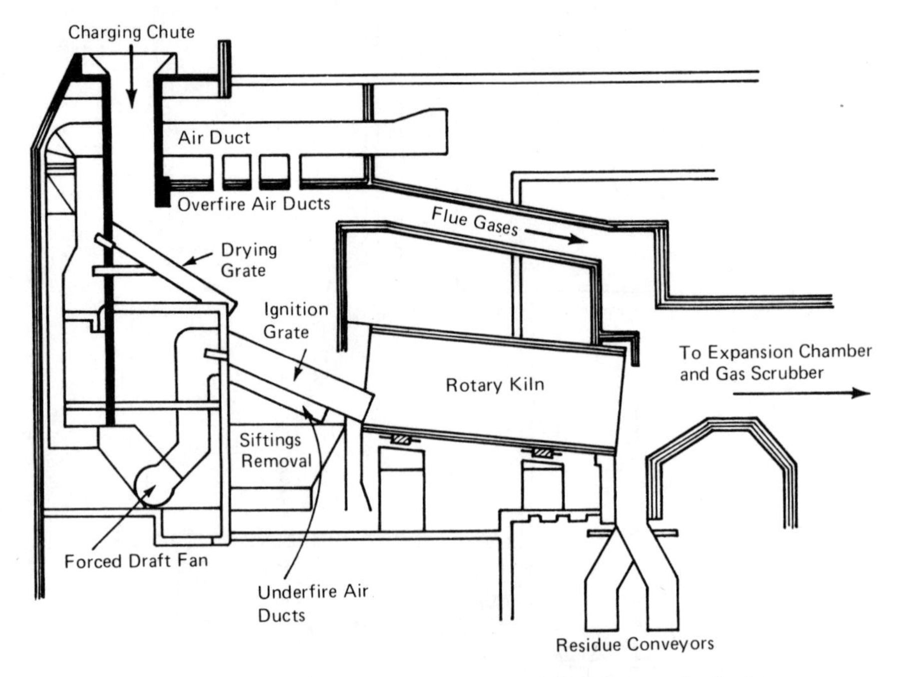

Figure 3-7. Rotary Kiln Furnace. Source: DeMarco *et al., Incinerator Guidelines–1969*, p. 30.

regard, a "Lexicon of Incinerator Terminology," by D. Schwartz, also in the *Proceedings of the National Incinerator Conference* (1964), may be of interest, especially to the layman. While this book has consciously avoided using highly specialized words, many readers will doubtless be aided by a glossary that covers such words as "tuyers" and "buckstays."

In spite of the many incinerator options, however, there are several advantages of continuous-feed systems that have made them the overwhelming choice for recent installations. C.A. Rogus, writing in *The American City,* has summarized these advantages as follows:

> In the larger installations, the continuously fed mechanically stoked furnace with continuous ash removal is replacing the batch-fed furnace with its manual or semi-mechanical stoking and its intermittent discharge of residue. The well-recognized inherent advantages of uniform burning at higher temperatures are control of excess air, more efficient and thorough combustion, less odor and air pollution, reduced slagging and clinkering, and alleviation of damage to refractories. Its economies in operation and maintenance and greater freedom from nuisances far outweigh the higher initial costs. If we add to this the problem of continually rising labor costs, the need for still greater

mechanization and near-automatic operation becomes more pressing.[1]

In addition to more highly automated, continuous-feed incineration plants, there is a trend toward larger plants, many designed for 24-hour operation. This concept will become increasingly important as waste heat recovery becomes more widespread. Since deliveries of refuse to the plant can normally be made only during about six daylight hours, storage bins must be provided in 24-hour plants to provide for continuous operations.

There are several new types of incinerator design now in various stages of development. These include slagging furnaces, suspension firing, fluidized bed incinerators, and pyrolysis units. Also, much development work is being done on methods for controlling air pollution from incinerators. Both of these areas are detailed later in this report.

Municipal Incineration in the United States

The best overall view of the status of municipal incineration in the United States is available in two reports: "Municipal Incinerator Design Practices and Trends," by J.W. Stephenson and A.S. Cafiero, in the *Proceedings of the National Incinerator Conference* (1966); and in *Systems Study of Air Pollution from Municipal Incineration,* by Walter Niessen, *et al.,* (Arthur D. Little, 1970). The 1966 work compiled a list of all known incinerators (that burn *municipal refuse*) that were built or rebuilt in the United States and Canada from 1945 through 1965. The list totaled 289 incinerators. To these operations, a comprehensive questionnaire was sent; there were 205 replies.

The 1970 work updated the 1966 work by surveying 364 municipal-refuse-burning incinerators that had been built or rebuilt from 1922 up to those under construction in 1969. While the latter survey is more comprehensive in areas such as air pollution control, the former survey covers some areas (such as the receiving pit) that are not covered at all in the 1970 work.

The following material comes from the two cited works. Table 3-1 presents the 364 additions, modifications, new units, and rebuilt units of the 1970 report. Note that some incinerators are on the list more than once if they were rebuilt or modified; for example, the Racine plant appears on the list three times (line 14, when the plant was first built in 1929; line 76, when the plant was rebuilt in 1946; line 169, when an addition was made to the plant). In addition to the plant location, the table shows the year in which the plant was built, the plant's status as of 1969, the plant's 24-hour capacity, the number of furnaces in the plant, type of draft equipment, type of furnace and grate, air pollution control equipment, and what, if any, by-products the plant produces. Note that as of 1969, 76 of the plants had been closed. Since 1969, a

Table 3–1
Incinerator Inventory

Line No.	Location	Year Built	1969 Status	Plant Capacity Tons Per 24 Hrs. (incl. add'ns)	Furnaces: No.	Furnaces: Capacity Ea. T/24 Hrs.	Furnaces: Type Furnace & Grate	Furnaces: Mech. Draft	APC Equipment	By-Products
1.	Philadelphia, Pa. (Harrowgate) (see line 165)	1922	Rebuilt '54	300	2	150	Batch/Stationary		None	Steam
2.	Minneapolis, Minn. (see line 45)	1924		150	1	150	Batch/Stationary		None	
3.	Hamilton Township, N.J.	1925	Add'ns/'42,'45	99	3	33	Hearth Type/Stationary		WW	
4.	Allentown, Pa. (see lines 23, 145)	1925		90	1	90	Conv. Boiler/Stationary		None	
5.	Laredo, Texas	1925	Closed '66	24	1	24	Hearth		None	
6.	Pennsauken, N.J. (see line 139)	1926	Rebuilt '53	60	1	60	Fixed Hearth/Circular		None	
7.	Hackensack, N.J.	1927	Closed	100	1	100	Stationary Grates		WW	
8.	Buffalo, N.Y.	1927	Closed '70	400	5	80	Stationary (Manual)		None	
9.	Tonawanda, N.Y. (Town) (see lines 90, 112)	1928		70	1	70	Batch Hearth/Manually Stoked			
10.	Middletown, N.Y.	1929	Closed '68	50	1	50	Stationary Grate		None	
11.	New Rochelle, N.Y.	1929	Closed '46	150	3	50	Hearth Type		None	
12.	Perth Amboy, N.J. (see line 172)	1929		50	1	50	Stationary		None	
13.	Oshkosh, Wisc.	1929		100	1	100	Hearth/Hand Stoked		None	
14.	Racine, Wisc. (see lines 76, 169, 233)	1929		120	2	60	Hearth		None	
15.	Red Bank, N.J. (see line 173)	1930		60	1	60	Hearth Type		None	
16.	Spring Lake, N.J.	1930		30	2	15	Stationary		None	
17.	Milwaukee, Wisc. (Erie St.)	1930		225	3	75	Rect. Cell/Pinhole Grates		None	
18.	Atlantic City, N.J.	1931	Closed '68	96	3	32	Stationary		None	
19.	Elmira, N.Y.	1931	Closed '68	100	2	50	Stationary		None	
20.	Princeton, N.J. (see line 157)	1932		60	2	30	Circular		None	
21.	Schenectady, N.Y.	1932	Closed '67	240	3	80	Stationary/Hand Stoked		None	
22.	Portland, Ore.	1932	Closed	75	3	25	Stationary		None	
23.	Allentown, Pa. (see lines 4, 145)	1932	Add'n	180	1	90	Stationary/Conv. Boiler		None	
24.	Washington, D.C. (Georgetown)	1932		170	2	85	Stationary Grate		WW	
25.	Washington, D.C. (O Street)	1932		425	5	85	Stationary Grate		None	
26.	Baltimore, Md.	1933		600	4	150	Rocking Grate		WW	

27.	Cincinnati, Ohio	1933		200	2	100	Rocking		None	
28.	Cincinnati, Ohio (Red Bank)	1933		200	2	100	Rocking		S. Ping Pong Ball Scrubber	
29.	Shreveport, La. (see lines 250, 99)	1934	Closed	150	1	150	Stationary		None	
30.	New York, N.Y. (Bronx)	1934	Closed	750	5	150	Stationary Dump Grate	U.O.	D	Salvage
31.	New York, N.Y. (W. 215th)	1934		750	5	150	Rect. Cell/Stat. Dump Grate	U.O.	D	Salvage
32.	Middletown, Conn.	1936	Closed	60	1	60	Hearth		None	
33.	Cleveland, Ohio	1936	Closed	900	3	300	Stationary		D	
34.	New York, N.Y. (Flushing)	1937	Closed	300	3	100	Stationary Dump Grates		None	
35.	Glen Cove, N.Y.	1937	Closed '69	100	2	50	Stationary Grates		None	
36.	Manchester, N.H.	1937		125			Rect. Cell/Flat Dump Gr.	U.O.I.D.	None	
37.	Greenwich, Conn. (see line 284)	1938		150	1	150	Batch-Rect./Contin.	Nat.	D.S.	
38.	Cambridge, Mass.	1938		150	3	50	Cell/Horiz./Hand Stoked	Nat.	None	
39.	Detroit, Mich. (St. Jean) (see line 212)	1938	Rebuilt '57	350	2	175	Stationary		None	
40.	Detroit, Mich. (NW) (see line 211)	1938	Rebuilt '57	450	3	150	Stationary		D	
41.	Detroit, Mich. (Central) (see lines 59, 228)	1938		350	2	175	Stationary		None	
42.	Detroit, Mich. (24th St.) (see line 290)	1938	Rebuilt '63	500	2	250	Batch/Stationary		D	
43.	Tonawanda, N.Y. (see line 229)	1938	Closed	50	1	50	Basket Type Furnace		None	
44.	Lower Merion, Pa. (see lines 94, 362)	1938	Closed '69	100	2	50	Stationary	Yes	None	
45.	Athens, Ga.	1939		48	1	48	Cylindrical		None	
46.	Minneapolis, Minn. (see line 2)	1939	Add'n	300	1	150	Stationary		D	
47.	Larchmont, N.Y.	1939	Closed '69	120	2	60	Batch/Stationary		D	
48.	New Rochelle, N.Y.	1939		250	2	125	Stationary/Manually Stoked		WWF	
49.	Pittsburgh, Pa.	1939		600	4	150	Batch/Stationary		None	
50.	Newport, R.I. (see line 314)	1939	Rebuilt '64	100	1	150	Manual		None	
51.	Providence, R.I. (see lines 95, 166)	1939		160	1	160	Batch/Man. Stoked/Stat. Dumping Grate	Yes	None	
52.	Durham, N.C.	1940	Closed	460	4	115	Stationary		D	
53.	Dayton, Ohio	1940		200			Cylindrical		None	
54.	New London, Conn.	1941		120			Rocking Grate		None	
55.	Atlanta, Ga. (see line 120)	1941		350	2	175	3 Grates/Rotary Kiln	U.O.I.	D	Steam, Salvage
56.	Stamford, Conn. (see line 235)	1942		225	3	75	Batch/Rocker Type		S.B.	
57.	Orlando, Florida	1942		150	2	75	Circular Mech.		None	
58.	Middletown, Conn.	1943	Closed	60	1	60	Mech. Stoker	U	None	
59.	Detroit, Mich. (Central) (see lines 41, 228)	1943	Add'n	525	1	175	Circular		None	
60.	Staunton, Va.	1944	Closed	60	1	60	Batch/Circular/Mech. Stoked		None	
61.	Jacksonville, Florida (S. Side)	1945	Closed	120	1	120	Batch/Mech. Stoked/Circular	Yes	None	

Table 3-1 (continued)

Line No.	Location	Year Built	1969 Status	Plant Capacity Tons Per 24 Hrs. (incl. add'ns)	Furnaces No.	Capacity Ea. T/24 Hrs.	Type Furnace & Grate	Mech. Draft	APC Equipment	By-Products
62.	Youngstown, Ohio	1945		200	2	100	Batch/Man. Stoked		None	
63.	Lexington, Virginia	1945	Closed '54	30	1	30			None	
64.	Fond du Lac, Wisc.	1945		90	3	30	Hearth		None	
65.	Kenosha, Wisc. (see line 283)	1945	Closed	300	2	150	Stationary		None	
66.	Bessemer, Ala.	1946	Closed	60	2	30	Batch/Man. Stoked		None	
67.	Beverly Hills, Calif.	1946	Closed '56	300	2	150	Batch/Circular/Mech. Stoked		None	
68.	Wash. Suburban Sanitary Comm., Maryland	1946	Closed '64	250	2	125	Batch/Circ. Mech. Stoked		None	
69.	Amsterdam, N.Y.	1946		120	2	60	Man. Stoked		None	
70.	Babylon, N.Y.	1946	Closed '55	90	1	90	Circular		None	
71.	Cheektowaga, N.Y.	1946		100	2	50	Batch/Mech. Stoked		None	
72.	Sidney, Ohio	1946	Closed	50	1	50	Batch/Circ./Mech. Stoked		None	
73.	Warwick, R.I.	1946		100	1	100	Batch/Mech. Stoked/Circ.		D	
74.	Norfolk, Va.	1946		400	4	100	Cylindrical		D	
75.	Green Bay, Wisc. (see line 296)	1946		50	1	50	Batch/Stationary		D	
76.	Racine, Wisc. (see lines 14, 169, 233)	1946	Rebuilt	220	{2 1	80 60	Batch/Man. Stoked	I	D	
77.	Jacksonville, Fla.	1947	Closed	350	2		Batch/Circ./Mech. Stoked		None	
78.	Aurora, Illinois	1947		40	2	20	Stationary		None	
79.	Holyoke, Mass.	1947		225	3	75	Rect. Cell/Manual Linkage Grate		D	
80.	Corning, N.Y.	1947		80	2	40	Batch/Circ./Mech. Stoked	U.O.	D	
81.	Troy, N.Y.	1947	Closed '58	250	2	125	Batch/Circ./Mech. Stoked		None	
82.	West Seneca, N.Y.	1947	Closed '61	60	1	60	Batch/Circ./Mech. Stoked		None	
83.	East Cleveland, Ohio	1947	Closed '63	100	1	100	Batch/Circ./Mech. Stoked		None	
84.	Houston, Texas (Velasco St.)	1947	Closed	300	2	150	Stationary/Cell		None	
85.	Jefferson Parish, La.	1948		90	1	90	Batch/Circ./Mech. Stoked	U	D	
86.	Fall River, Mass.	1948	Closed	85	1	85	Hearth		None	
87.	Pittsfield, Mass.	1948		180	2	90	Batch/Man. Stoked		D	
88.	Carmel, N.Y.	1948		40	1	40	Batch/Circ./Mech. Stoked	U	None	
89.	Liberty, N.Y.	1948	Closed '68	30	1	30	Batch/Man. Stoked		None	
90.	Tonawanda, N.Y. (Town) (see lines 9, 112)	1948	Add'n	140	2	70	Batch/Circ./Mech. Stoked		None	
91.	Barberton, Ohio	1948		100	2	50	Hearth		None	
92.	Cleveland Heights, Ohio	1948		150	2	75	Batch/Circ./Mech. Stoked	U	D	
93.	Columbus, Ohio	1948	Closed	150	3	50	Batch/Circ./Mech. Stoked		None	

No.	Location	Year	Status	Capacity	Units	Unit capacity	Type			
94.	Lower Merion, Pa. (see lines 44, 362)	1948	Add'n	150	2	75	Rocking Grate Installed		None	
95.	Providence, R.I. (see lines 51, 166)	1948	Add'n	320	1	160	Batch/Double Traveling Grate		None	Steam
96.	Monroe, Wisc. (see line 297)	1948		30	1	30	Man. Stoked		None	
97.	West Bend, Wisc.	1948	Closed	30	1	30	Batch/Circ./Mech. Stoked		None	
98.	Morgan City, La.	1949		30	1	30	Batch/Man. Stoked		None	
99.	Shreveport, La. (see lines 29, 250)	1949	Add'n	175	1	175	Stationary		None	Changed
100.	Salisbury, Md.	1949		125	1	125	Circ./Mech. Stoked		F	
101.	Lackawanna, N.Y.	1949		150	2	75	Stat. Man. Stoked/ Dumping Grate		None	
102.	Mt. Kisco, N.Y.	1949		40	1	40	Batch/Man. Stoked	U	None	
103.	Mt. Vernon, N.Y.	1949		600	4	150	Batch/Circ./Mech. Stoked	U	None	
104.	Collingswood, N.J.	1949	Closed	40	1	40	Batch/Circ./Mech. Stoked		None	
105.	Meadville, Pa.	1949		80	1	80	Batch/Circ./Mech. Stoked	U	D	
106.	West Mifflin, Pa.	1949		40	3	1/20 2/10	Stationary		None	
107.	Houston, Texas (Patterson) (see line 167)	1949	Closed	300	2	150	Cell/Stationary		None	
108.	Alhambra, Calif.	1950	Closed '61	150	1	150	Batch/Circ./Mech. Stoked	U.I	None	
109.	Jacksonville, Fla.	1950		300	2	150	Batch/Circ./Mech. Stoked	SB		
110.	Gretna, La.	1950		100	2	50	Batch/Circ./Mech. Stoked	U	D	
111.	Gloucester City, N.J.	1950	Closed '64	60	1	60	Batch/Circ./Mech. Stoked		S	
112.	Tonawanda, N.Y. (Town) (see lines 9, 90)	1950	Add'n	200	1	60	Batch/Circ./Mech. Stoked		None	
113.	Ambridge, Pa.	1950	Closed	150	2	75	Stationary		None	
114.	Philadelphia, Pa. (Southeast) (see line 179)	1950		300	2	150	Batch/Circ./Mech. Stoked		D–SB	
115.	Philadelphia, Pa. (Bartram)	1950		300	2	150	Batch/Circ./Mech. Stoked		D. WW	
116.	Arlington, Va. (see lines 216, 330)	1950		300	2	150	Batch/Mech. Stoked		WW	
117.	Pomona, Calif.	1951	Closed '55	225	2	112.5	Batch/Circ./Mech. Stoked/MB		None	
118.	Derby, Conn.	1951	Closed	60	1	60	Circ.		None	
119.	Miami, Fla.	1951		900	6	150	Cont./Circ./Mech. Stoked	I	WW	Steam
120.	Atlanta, Ga. (Mayson) (see line 55)	1951	Add'n	350	2	175	3 Grates/Rotary Kiln	U.O.I.	D	
121.	St. Louis, Mo. (South Bend)	1951		400	4	100	Batch/Rocking Grate	Yes	D	
122.	Port Chester, N.Y.	1951		120	2	60	Batch/Circ./Mech. Stoked	O	D. F.	Steam
123.	Yonkers, N.Y.	1951		450	3	150	Batch/Circ./Mech. Stoked	Yes	D	
124.	Lakewood, Ohio	1951		150	2	75	Batch/Man. Stoked	Yes	D	
125.	Lima, Ohio	1951	Closed	200	2	100	Batch/Stationary		None	
126.	Warren, Ohio		Closed '67	100	3	33+	Batch/Shaker		None	
127.	Glendale, Calif. (see line 185)	1952	Closed	90	1	90	Batch/Circ./Mech. Stoked		None	
128.	Santa Monica, Calif.	1952	Closed	300	4	75	Batch/Man. Stoked	U.O.I.	None	
129.	Waterbury, Conn.	1952		300	2	150	Batch/Circ./Mech. Stoked	U	WW	Drying, Sewage, S
130.	Brookline, Mass.	1952		300	2	150	Batch/Mech. Stoked/ Rabble Arm	Yes	D, SB	

Table 3-1 (continued)

Line No.	Location	Year Built	1969 Status	Plant Capacity Tons Per 24 Hrs. (incl. add'ns)	Furnaces				APC Equipment	By-Products
					No.	Capacity Ea. T/24 Hrs.	Type Furnace & Grate	Mech. Draft		
131.	Lawrence, Mass.	1952		300	2	150	Batch/Circ./Mech. Stoked	I	C	Steam
132.	Somerville, Mass.	1952		450	3	150	Batch/Circ./Mech. Stoked		SB	
133.	Hempstead, N.Y.	1952		700	4	175	Batch/Circ./Mech. Stoked	I	C.WB	Steam
134.	North Hempstead, N.Y.	1952		200	2	100	Batch/Circ./Mech. Stoked		S.C	
135.	Rocky River, Ohio	1952	Closed	50	1	50	Batch/Hearth		None	
136.	Bloomsburg, Pa.	1952	Closed '66	60	1	60	Batch/Circ./Mech. Stoked	Yes	None	Steam
137.	Los Angeles, Cal. (Gaffey St.)	1953	Closed	200	2	100	Batch/Circ./Mech. Stoked	Yes	D	
138.	S.E. Oakland Co., Mich. (see line 291)	1953		450	3	150	Batch/Circ./Mech. Stoked	Yes	D	
139.	Pennsauken, N.J. (see line 6)	1953	Closed	60	1	60	Batch/Circ./Mech. Stoked		S	
140.	Woodbridge, N.J.	1953	Closed '62	300	2	150	Batch/Stationary/Drop Gate		None	
141.	Harrison, N.Y.	1953		150	2	75	Batch/Circ./Mech. Stoked	U	F	
142.	Long Beach, N.Y.	1953		200	2	100	Batch/Circ./Mech. Stoked		EB	
143.	New York, N.Y. (Gansevoort)	1953		1000	4	250	Cont./Double Travel. Grate		D.WW	
144.	Cheviot, Ohio	1953		60	2	30	Batch/Stationary		None	
145.	Allentown, Pa. (see lines 4, 23)	1953	Add'n	270	1	90	Stationary/Conv. Boiler		None	
146.	Erie, Pa.	1953		100	2	50	Batch/Circ./Mech. Stoked		D	
147.	Fort Worth, Texas	1953	Closed	260	2	130	Batch/Circ./Mech. Stoked		None	
148.	Milwaukee, Wisc. (Green Bay Ave.)	1953		300	2	150	Batch/Circ./Mech. Stoked	U.O.	WW	Steam
149.	Hartford, Conn.	1954		600	4	150	Circ./Mech. Stoked	U	D	
150.	New Britain, Conn.	1954		300	2	150	Cylindrical Grate		None	
151.	Ft. Lauderdale, Fla.	1954		250	2	125	Circ./Mech. Stoked		None	
152.	Skokie, Ill.	1954		150	2	75	Batch/Impact Stoker	Nat.	WW	
153.	Newton, Mass. (see line 351)	1954		240	2	120	Batch/Rocking Grate		SB	
154.	Ecorse, Mich.	1954	Closed	90	1	90	Batch/Circ./Mech. Stoked		None	
155.	St. Louis Park, Minn.	1954	Closed	150	2	75	Batch/Rocking Grate	Yes	None	
156.	Omaha, Neb.	1954		375	3	125	Batch/Cell Type/Rocking Grate	U.O.	D	
157.	Princeton, N.J. (see line 20)	1954	Add'n	120	1	60	Batch/Circ.		None	
158.	Buffalo, N.Y.	1954		600	3	200	Batch/Circ./Mech. Stoked	Yes	D	
159.	New York, N.Y. (S. Shore)	1954		1000	4	250	Cont./Double Trav. Grate	U.O.	D.F	
160.	Huntington, N.Y.	1954		150	1	150	Batch/Rocking Gr./Man. Oper.	Yes	S	
161.	Berea, Ohio	1954	Closed '66	50	1	50	Batch/Man. Stoked		None	
162.	Cincinnati, Ohio (West Fork)	1954		500	4	125	Batch/Circ./Mech. Stoked		D	
163.	South Euclid, Ohio	1954	Closed	100	2	50	Batch/Rocking Grate F & E	U	None	
164.	Youngstown, Ohio	1954	Add'n	100	1	100	Batch/Man. Stoked	Yes, I	D.Sc	

165.	Philadelphia, Pa. (Harrowgate) (see line 1)	1954	Rebuilt	300	2	150	Batch/Circ./Mech. Stoked		D.WW	
166.	Providence, R.I. (see lines 51, 95)	1954	Rebuilt	320	1	160	Cont./Double Trav. Grate	Yes, I	None	Steam
167.	Houston, Texas (Patterson) (see line 107)	1954	Closed	300	2	150	Cell/Stationary Grate		None	
168.	Kenosha, Wisc.	1954	Closed	120	2	60	Manual, Stationary		None	
169.	Racine, Wisc. (see lines 14, 76, 233)	1954	Add'n	60	1	60	Rocking Gr. F & E	I.U.	D.WW	
170.	Wash., D.C. (#3 Mt. Olivet)	1954		500	4	125	Batch/Rocking Grate	U	D	
171.	Framingham, Mass.	1955		200	2	100	Cont./Single Traveling Grate	U.I.	S.Sc	
172.	Perth Amboy, N.J. (see line 12)	1955	Add'n	150	1	100	Stationary		None	
173.	Red Bank, N.J. (see line 15)	1955	Add'n	120	1	60	Hearth		WW	
174.	Babylon, N.Y.	1955		300	2	150	Batch/Circ./Mech. Stoked		S	
175.	Rochester, N.Y. (West Side)	1955		450	3	150	Batch/Circ./Mech. Stoked		WW.F	
176.	Bedford, Ohio	1955	Closed	50	1	50	Dump/Stationary		None	
177.	Abington, Pa.	1955		200	2	100	Batch/Rocking Grate	U.O.	D	
178.	Philadelphia, Pa. (Bartram) (see line 115)	1955	Add'n	450	2	125	Batch/Circ./Mech. Stoked		D.WW	
179.	Philadelphia, Pa. (Southeast) (see line 114)	1955	Add'n	300	2	150	Batch/2-Trav.		WW	
180.	Red Lion Borough, Pa.	1955		60	1	60	Batch/Circ./Mech. Stoked		WW	
181.	Houston, Texas (Holmes Rd.)	1955	Closed	300	1	300	Cell-Stationary		None	
182.	Kewaskum, Wisconsin	1955		24	1	24	Batch/Circ./Mech. Stoked		WW	
183.	Merrill, Wisc.	1955		35	1	35	Rocking Grate	Yes	D	
184.	Milwaukee, Wisc. (Lincoln Ave.)	1955		300	2	150	Batch/Rocking Grate	U.O.	WW	Steam
185.	Glendale, Calif. (see line 127)	1956	(Add'n) Closed	180	1	90	Batch/Ram Feed/Rect. Grate	U.I.	WB.S	
186.	Los Angeles, Calif. (Lacy St.)	1956	Closed	320	2	160	Batch/Man. Stoked	O.I.	D.WW	
187.	East Hartford, Conn. (see line 333)	1956		200	2	100	Batch/Rocking Grate	Nat.	D	
188.	New Canaan, Conn.	1956	Rebuilt	50	1	50	Batch/Circ./Mech. Stoked	U	D	
189.	West Hartford, Conn.	1956		350	2	175	Batch/Mech. Stoked/Circ.	U	D	Steam
190.	Chicago, Ill. (Medill)	1956		720	4	180	Batch/Rocking Grate	Yes	S	Steam
191.	Baltimore, Md. (#4)	1956		800	4	200	Batch/Rocking Grate	U.O.	D	
192.	St. Louis, Mo. (North Side)	1956		400	4	100	Rocking Grate/F & E/Rect. Cell	Yes	D	
193.	Binghamton, N.Y.	1956		300	2	150	Batch/Circ./Mech. Stoked	U	D	
194.	Niagara Falls, N.Y.	1956		240	2	120	Batch/Circ./Mech. Stoked	Yes	D	
195.	Oyster Bay, N.Y. (Bethpage)	1956		500	4	125	Batch/Rect. Cell	U.I.	SB	Steam
196.	Poughkeepsie, N.Y.	1956	Rebuilt	200	2	100	Batch/Circ./Mech. Stoked	U	D	
197.	Rochester, N.Y. (East Side)	1956		600	4	150	Batch/Circ./Mech. Stoked		WW.F	
198.	White Plains, N.Y.	1956		400	2	200	Batch/Rocking Grate	U	WW	
199.	Euclid, Ohio	1956		200	2	100	Batch/Rocking Grate	U.O.	WB	
200.	Maple Hts., Ohio	1956	Closed '67	150	2	75	Batch/Rocking Grate		None	
201.	Parma, Ohio	1956		225	2	112.5	Batch/Rocking Grate	U	SB	Hot Water

Table 3–1 (continued)

Line No.	Location	Year Built	1969 Status	Plant Capacity Tons Per 24 Hrs. (incl. add'ns)	Furnaces No.	Furnaces Capacity Ea. T/24 Hrs.	Furnaces Type Furnace & Grate	Furnaces Mech. Draft	APC Equipment	By-Products
202.	Philadelphia, Pa. (Northeast)	1956		600	4	150	Batch/Circ./Mech. Stoked	U.I.	D.WW	
203.	Fort Worth, Texas	1956	Closed	190	2	95	Batch/Circ./Mech. S		None	
204.	Alexandria, Va.	1956	Standby	200	2	100	Cell Type/Rocking Grate	Yes	D	
205.	West Allis, Wisc.	1956		200	2	100	Batch/Rocking Grate		S.EB	
206.	Whitefish Bay, Wisc.	1956		80	2	40	Cont./Rect. Cell		D.WW	
207.	Coral Gables, Fla.	1957		300	2	150	Batch/Circ./Mech. Stoked		WW	
208.	Lexington, Ky.	1957		200	2	100	Batch/Circ./Mech. Stoked	Yes	D	
209.	Louisville, Ky. (see line 321)	1957		750	3	250	Cont. Rot. Kiln		S	Steam, Salvage Cans
210.	Worcester, Mass.	1957		450	3	150	Batch/Circ./Mech. Stoked		None	
211.	Detroit, Mich. (Northwest) (see line 40)	1957	Rebuilt	450	2	225	Batch/Rocking Grate		WW	
212.	Detroit, Mich. (St. Jean) (see line 39)	1957	Rebuilt	350	2	175	Batch/Rocking Grate		WW	
213.	Hamtramck, Mich.	1957	One closed '66	100	2	100	Rocking Grate		S	
214.	Jersey City, N.J.	1957		600	4	150	Batch/Circ./Mech. Stoked	U	Sc	Steam
215.	New York, N.Y. (74th St.)	1957	Rebuilt	660	3	220	Cont/Double Traveling Gr.	U.O.I.	C	Steam
216.	Arlington, Va. (see lines 116, 330)	1957	Add'n	600	2	150	Batch/Circ./Mech. Stoked		SB	
217.	Bridgeport, Conn.	1958		300	2	150	Batch/Circ./Mech. Stoked	U.O.	W.S.SB	
218.	New Milford, Conn.	1958	Closed	75	1	75	Rolling Grate		None	
219.	Hollywood, Fla.	1958		450	2	225	Batch/Oscillating Grate	Yes	WB.S	
220.	Evanston, Ill.	1958		180	2	90	Batch/Rocking Grate	Yes	D	
221.	Melrose Park, Ill.	1958		250	1	250	Batch/Impact		None	
222.	Schiller Park, Ill.	1958		250	1	250	Modified Internatl.		None	
223.	Stickney, Ill.	1958		500	2	250	Cont./Grate & Kiln		WB	
224.	Indianapolis, Indiana	1958		450	3	150	Batch/Circ./Mech. Stoked		WW	Steam
225.	New Orleans, La. (Fla. Ave.)	1958		400	2	200	Batch/Rocking Grate	U	D.S	
226.	Marblehead, Mass.	1958		90	1	90	Batch/Rocking Grate	U	WW	
227.	Watertown, Mass.	1958		250	2	125	Batch/Circ./Mech. Stoked	I.D.O.U.	Sc.C	
228.	Detroit, Mich. (Central) (see lines 41, 59)	1958	Modification	525	3	175	Batch/Rocking Grate		WW	
229.	Tonawanda, N.Y. (see line 43)	1958	Add'n	100	2	50	Rocking Grate		D	
230.	Durham, N.C.	1958	Closed	300	2	150	Batch/Man. Stoked/ Hydraulic Dump	U	D	
231.	Fort Worth, Texas	1958	Closed	125	2	62.5	Batch/Rocking Grate		None	

232.	Neenah-Menasha, Wisc.	1958		300	2	150	Cont./Single Trav. Grate		S	
233.	Racine, Wisc. (see lines 14, 76, 169)	1958	Add'n	120	1	60	Rocking Grate	I.U.	D.WW	
234.	Shorewood, Wisconsin	1958		60	2	30	Batch/Rocking Grate	U.O.	WW	
235.	Stamford, Conn. (see line 56)	1959	Add'n	125	1	125	Batch/Rocking Grate	U	SB	Steam
236.	Chicago, Ill.	1959		1200	6	200	Batch/Rocking Grate		WW	Steam
237.	New Albany, Indiana	1959		160	2	80	Cont./Single Trav. Grate	U.O.	D	Steam
238.	Belmont, Mass.	1959		150	2	75	Batch/Circ./Mech. Stoked	U.I.O.	SB	Salvage
239.	Boston, Mass.	1959		900	6	150	Batch/Rocking Grate	U.I.	C	Steam
240.	New Bedford, Mass.	1959		225	2	112.5	Rocking Grate		None	
241.	Waltham, Mass.	1959		200	2	100	Rocking Grate	Nat.	D	
242.	Wellesley, Mass.	1959		150	2	75	Batch/Rocking Grate		F	
243.	New York, N.Y. (Betts Ave.)	1959	Rebuilt	1000	4	250	Cont./Double Trav. Grate	U.O.	D.F	
244.	New York, N.Y. (Greenpoint)	1959		1000	4	250	Cont./Double Trav. Grate	U.O.	None	Steam
245.	Rye, N.Y.	1959		150	2	75	Batch/Circ./Mech. Stoked	U	WB	
246.	Scarsdale, N.Y.	1959		150	2	75	Batch/Circ./Mech. Stoked	U	WB	
247.	Whitemarsh, Pa. (see line 356)	1959		300	1	300	Batch/Charging Ram/Recip. Grate	U.O.I.	Sc	
248.	Miami, Fla.	1960		300	1	300	Cont./Grates and Rotary Kiln	U.O.	SB	
249.	Winchester, Kentucky	1960		100	2	50	Batch/Reciprocating Grate	I	SB	
250.	Shreveport, La. (see lines 29, 99)	1960	Rebuilt	250	1	250	Rocking Grate		S	
251.	Garden City, Mich.	1960		500					WW	
252.	Huntington, N.Y.	1960		150	1	150	Batch/Rocking Gr/Man. Oper.		None	
253.	New Rochelle, N.Y.	1960		150	1	150	Batch/Rocking Grate		WW.F	
254.	North Tonawanda, N.Y.	1960	Closed '69	100	1	100	Batch/Reciproc.		None	
255.	Cleveland, Ohio	1960		500	4	125	Batch/Rocking Grate	I	WW	
256.	Bradford, Pa.	1960		200	2	100	Batch/Rocking Grate	U.O.	D	Steam
257.	Delaware County, Pa. (#1)	1960		500	2	250	Cont./Double Trav. Grate	U.O.	D.F	
258.	Philadelphia, Pa. (NW)	1960		600	2	300	Cont./Double Trav. Grate	U.O.I.	S.C	
259.	Woonsocket, R.I.	1960		160	2	80	Batch/Circ./Mech. Stoked	U	D	
260.	Bridgeport, Conn.	1961		200	2	100	Batch/Rocking Grate		S	
261.	Honolulu, Hawaii	1961		220	2	110	Batch/Recipr. Grate		WB	
262.	Dedham, Mass.	1961		100	2	50	Batch/Rocking Grate	U.O	SB	
263.	Winchester, Mass.	1961		100	2	50	Batch/Rocking Grate	U.O.I.	WB.S.C.	
264.	River Rouge, Mich.	1961		60	1	60	Batch/Rocking Grate		S.WW	
265.	New York, N.Y. (Hamilton Ave.)	1961		1000	4	250	Cont./Double Traveling Grate	U.O	SB.D	Steam
266.	New York, N.Y. (S.W. Brooklyn)	1961		1000	4	250	Cont./Double Traveling Grate	U.O.	S	
267.	Norwood, Ohio	1961		150	2	75	Batch/Recipr. Grate	U.O.I.	S	
268.	Sharonville, Ohio (see line 340)	1961		350	1	350	Cont./Double Trav. Grate	I	SC	
269.	Delaware County, Pa. (#2)	1961		500	2	250	Cont./Double Trav. Grate		WB	
270.	DePere, Wisc. (see line 344)	1961		225	1	150	Hearth	Yes	D.S	
					1	75	Rect. Cell/Rocking Grate	Yes	D.S	
271.	Wauwatosa, Wisc.	1961		165	2	82.5	Batch/Rocking Grate	Yes	WW	

Table 3-1 (continued)

Line No.	Location	Year Built	1969 Status	Plant Capacity Tons Per 24 Hrs. (incl. add'ns)	Furnaces				APC Equipment	By-Products
					No.	Capacity Ea. T/24 Hrs.	Type Furnace & Grate	Mech. Draft		
272.	Washington, D.C. (Ft. Totten)	1961		500	4	125	Batch/Rocking Grate	U	D	
273.	Darien, Conn.		(Add'n)							
		1962	Closed	70	1	70	Batch/Rocking Grate	U	WW	
274.	Norwalk, Conn.	1962		360	2	180	Cont/Single Trav. Grate	U.O	SB	
275.	Honolulu, Hawaii	1962		220	2	110	Batch/Reciprocating Grate		Sc.WW	
276.	New Orleans, La. (7th St.)	1962		400	2	200	Batch/Rocking Grate	U	D.S	
277.	Salem, Mass.	1962		230	2	115	Batch/Circ.	Nat.	SB	
278.	Eastchester, N.Y.	1962		200	2	100	Batch/Rocking Grate	U	WB.S	
279.	Islip, N.Y.	1962		300	2	150	Cont./Single Trav. Grate	U.O.I.	C	
280.	Valley Stream, N.Y.	1962		200	2	100	Cont./Single Trav. Grate	U.O.	WW	
281.	Woodville, Ohio	1962		12	1	12	Batch/Man. Stoked		None	
282.	Delaware County, Pa. (#3)	1962		500	2	250	Cont./Double Trav. Grate	U.O.	WB.SB	
283.	Kenosha, Wisc.	1962	Closed	220	{2	50				
	(see line 65)				{2	60	Shaker		None	
284.	Greenwich, Conn. (see line 37)	1963	Add'n	400	1	250	Cont./Rocking Grate	U.O.	SB	
285.	New Haven, Conn.	1963		720	3	240	Cont./Double Trav. Grate	U.O.S.	WB	
286.	Atlanta, Ga.	1963		500	2	250	Cont./3 Grates & Kiln	U.O.	WB.S.SB	Metals, Rotat?
287.	Chicago, Ill. (SW)	1963		1200	4	300	Rotary Kiln/Recip. Grate	U.I.	WB.S.SB	Steam
288.	New Orleans, La. (Algiers)	1963		200	1	200	Cont./Double Trav. Grate	U.O.S.	C	
289.	Watertown, Mass.	1963		100	2	50	Circ.	I	S. Multiple-Cyclone	
290.	Detroit, Mich. (24th St.) (see line 42)	1963	Rebuilt	500	2	250	Batch/Rocking Grate		WW	
291.	S.E. Oakland Co., Mich. (see line 138)	1963	Rebuilt	600	2	300	Cont./Double Trav. Grate	U.O.	SB	Steam
292.	Garden City, N.Y.	1963		175	2	87.5	Cont./Recipr.	U.O.	WW	Steam
293.	Rocky River, Ohio	1963	Closed	60	1	60	Batch/Reciprocating Grate		D.S	
294.	Ambridge, Pa.	1963		150	2	75	Rocking Grate		None	
295.	Portsmouth, Virginia	1963		350	2	175	Batch/Rocking Grate	Yes	WW	
296.	Green Bay, Wisc. (see line 75)	1963	Add'n	150	2	50	Batch/Stationary		D	
297.	Monroe, Wisc. (see line 96)	1963	Add'n	60	1	30	Batch/Man. Stoked		None	
298.	Broward City, Fla.	1964		300	2	150	Cont./Recipr. Grate	U.O.I.	Sc	Metal
299.	Clearwater, Fla.	1964		300	2	150	Cont./Recipr. Grate	U.O.I.	WW	

300.	Orlando, Fla.	1964		250	2	125	Batch/Oscillating Grate	U.O.	SB	
301.	DeKalb County, Ga.	1964		600	2	300	Cont./Rotary Kiln	U.O.	S	
302.	Bloomington, Ind.	1964		100	1	100	Batch/Recipr. Grate	U.I.	SB	
303.	Frankfort, Kentucky	1964		150	2	75	Batch/Reciprocating Grate	I	SB	
304.	Jefferson Parish, La.	1964		400	2	200	Cont./Double Trav. Grate	U.O.S.	C	
305.	Lowell, Mass.	1964		400	2	200	Cont./Double Trav. Grate	O.I.U.	S	
306.	Central Wayne Co., Mich.	1964		500	2	250	Recipr.		SB	
307.	Dearborn, Mich.	1964		500	2	250	Cont./Recipr. Grate	U.O.I.	SB	
308.	Trenton, Mich.	1964		100	1	100	Batch/Recipr. Grate		S	
309.	Beacon, N.Y.	1964		100	2	50	Batch/Rocking Grate	U.O.	SB	
310.	Canajoharie, N.Y.	1964		50	1	50	Batch./Circ./Mech. Stoked	U	SB	
311.	Freeport, N.Y.	1964	Rebuilt	150	2	75	Batch/Rock. Grate-Constant Flow	U.O.	WB	
312.	Cincinnati, Ohio (Center Hill)	1964		500	2	250	Cont./Double Trav. Grate	U.O.	WW	
313.	Ewing Township, Pa.	1964		240	1	240	Cont./Single Trav. Grate	U.O.	D	
314.	Newport, R.I. (see line 50)	1964		120	1	120	Rocking Grate		None	
315.	Pawtucket, R.I.	1964		400	2	200	Cont./Double Trav. Grate		WW	
316.	Roanoke, Va.	1964	Rebuilt	200	2	100	Batch/Reciprocating Grate	U.O.	D	
317.	Charleston, W. Va.	1964		300	2	150	Batch/Reciprocating Grate	U.O.I.	SB	
318.	Nekoosa, Wisc.	1964		60	1	60	Batch/Reciprocating Grate	U	WB.S	
319.	Broward Cty., Fla.	1965		300	2	150	Cont./Recipr. Grate	U.I.O.	Sc	Metal
320.	Dodge City, Kansas	1965		35	1	35	Hearth		None	
321.	Louisville, Kentucky (see line 209)	1965	Add'n	1000	1	250	Cont./Rotary Kiln	Yes	S	Steam, Cans
322.	Paris, Kentucky	1965		100	2	50	Batch/Reciprocating Grate		SB	
323.	Montgomery County, Md.	1965		1050	3	350	Cont./Double Trav. Grate	U.O.I.	C	Steam
324.	Weymouth, Mass.	1965		300	2	150	Batch/Circ./Mech. Stoked	U	SB	
325.	Ewing, N.J.	1965		240	1	240	Single Trav./Horiz.	O.U.	D	
326.	Newburgh, N.Y.	1965		240	2	120	Cont./Rock. Grate-Const. Flow	U.O.S.	SB	
327.	Oyster Bay, N.Y.	1965		500	2	250	Cont./Rock. Grate-Const. Flow	U.O.	SB	
328.	Ramapo, N.Y.	1965	Add'n	200	1	200	Batch/Rect.	Yes	SB	
329.	Amarillo, Texas	1965		250	2	125	Batch/Ram Feed/Recipr. Grate		WW	
330.	Arlington, Va. (see lines 116, 216)	1965	Add'n	750	1	150	Circ./Mech. Stoked		SB	
331.	Port Washington, Wisc.	1965		75	1	75	Batch/Recipr. Grate	U.I.	SB	
332.	Sheboygan, Wisc.	1965		240	2	120	Cont./Rock. Gr./Const. Flow	U.O.	S	
333.	E. Hartford, Conn. (see line 187)	1966	Add'n	350	1	150	Batch/Rock. Grate	I.O.	D.SB	
334.	Ft. Lauderdale, Fla.	1966		450	2	225	Cont./Recipr. Grate	U.O.I.	S	
335.	Picayune, Miss.	1966		144	1	144			WW	
336.	Babylon, N.Y.	1966		400	2	200	Cont./Rock. Gr./Const. Flow	U.O.	S	
337.	Hempstead, N.Y. (see line 187)	1966		750	1	150	Cont./Rock. Grate/Const.	U.O.I.	SB	
					2	300				

Table 3-1 (continued)

Line No.	*Location*	*Year Built*	*1969 Status*	*Plant Capacity Tons Per 24 Hrs. (incl. add'ns)*	*Furnaces*				*APC Equipment*	*By-Products*
					No.	*Capacity Ea. T/24 Hrs.*	*Type Furnace & Grate*	*Mech. Draft*		
338.	Huntington, N.Y.	1966		150	1	150	Cont./Rock. Gr./Const. Flow	U.O.	D.Sc	
339.	North Hempstead, N.Y.	1966		600	2	200	Rock. Grate-Const. Flow	I	C	
					1	200	Double Trav. Grate			
340.	Sharonville, Ohio (see 268)	1966	Add'n	300	2	150	Batch/Recipr. Grate		WB	
341.	Philadelphia, Pa. (E. Central)	1966		600	2	300	Cont./Double Trav. Grate	I.O.U.	SB.WB	
342.	Ogden, Utah	1966		300	2	150	Single Traveling Grate		S	
343.	Alexandria, Va. (#2)	1966		300	2	150	Cont./Rock. Gr. Const. Flow	U.O.	D.SB	
344.	DePere, Wisc. (see line 270)	1966		300	1	300	Hearth		S	
345.	Green Bay, Wisc.	1966		200	2	100	Batch/Stationary		SB	
346.	West Haven, Conn.	1967		300	2	150	Cont./Rock. Gr.-Const. Flow	U.O.I.	S.C	
347.	St. Petersburg, Fla.	1967		500	2	250	Batch/Rock. Gr.-Const. Flow	U.O.I.	SB	
348.	Tampa, Fla.	1967		900	3	300	Cont./Kiln	U.O.I.	Sc	
349.	Lexington, Ky.	1967		150	1	150	Batch/Rocking Grate	U.O.	D	
350.	New Orleans, La. (East)	1967		400	2	200	Cont./Rock. or Recipr. Grate	U.O.S.	D.C	
351.	Newton, Mass. (see line 153)	1967		500	2	250	Cont./Triple Trav. Grate	U.O.	WW	
352:	Troy, Ohio	1967		150	1	150	Batch/Recipr.		SB	
353.	Houston, Texas (Holmes Rd.)	1967		800	2	400	Cont./Triple Trav. Gr.		Sc	
354.	Stratford, Conn.	1968		240	2	120	Cont./Single Trav. Grate	U.O.S.	SB	
355.	Miami County, Ohio	1968		150	1	150	Cont./3 Pusher Grates	I.U.O.	Sc	
356.	Whitemarsh, Pa. (see line 247)	1968	Rebuilt	300	1	300	Cont./Rocking Grate	U.O.I.	SB	
357.	Norfolk, Va.	1968		360	2	180	Cont./Recipr. Grate	U.O.I.	C. Waterwall Boiler	
358.	Oshkosh, Wisc.	1968		350	2	175	Cont./Recipr./Triple Grate	I.O.U.	S	
359.	Dade County, Fla.	1969 (est.)		300			Reciprocating		Sc	
360.	East Chicago, Ind.	1969 (est.)		450			Reciprocating		Sc	
361.	New Orleans, La. (7th St.)	1969	Rebuilt	400	2	200	Cont./Double Trav. Gr.	U.O.I.	S.C	
362.	Lower Merion, Pa. (see lines 44, 94)	1969		250	2	125	Cont./Rock. Gr./Const. Flow		Sc	
363.	Braintree, Mass.	1970 (est.)		480	2	240				Steam
364.	Fall River, Mass.	U.C.		600	2	300	Double Grate/Rock. or Recipr.		C	

Key to Abbreviations

Mechanical Draft

Yes – Forced Draft reported, but distribution not indicated.
U – Forced underfire. Includes cone cooling air for circular furnaces.
O – Forced overfire.
S – Forced side fire.
I – Induced draft.

Fly Ash Removal

C – Cyclones.
D – Dry expansion chamber.
F – Fly ash screen.
S – Water sprays.
SB – Spray or wet baffles (close-spaced baffles).
Sc – Scrubber.
WB – Water bottoms or ponds in chambers.
WW – Wet walls (transverse, wide-spaced baffles).

Source: Niessen *et al.*, Arthur D. Little, Inc., *Systems Study of Air Pollution from Municipal Incineration,* Volume II, Appendix I, pp. I5–I23.

number of additional plants such as the Mason plant in Atlanta, Georgia, and the plant in Miami, Florida, have been closed, primarily because the incinerator was not able to meet newer, more stringent air pollution control laws.

Figure 3-8 presents the total annual refuse tonnage of new incineration capacity that has been built since 1930. As can be seen from the figure, the major portion of the nation's incinerator capacity was built between 1950 and the early 1960s. Since 1965, the rate at which incinerators have been built has declined. Table 3-2 presents a breakdown, by stoker type, of the 316 plants that have been built in the United States between 1940 and 1970. Figure 3-9 is a plot of the average furnace capacity as a function of the year in which the furnace was built. Note that there has been a fairly steady trend toward larger furnaces. Figure 3-10 is a bar chart of the type of furnace built versus the year in which the furnace was built. As can be seen, just as the average furnace capacity has increased, the trend has been from batch-fed units to continuous-feed units.

Figure 3-11 is a bar chart of primary air pollution control equipment versus the year in which the plant was built. Note that the trend has been toward more sophisticated air pollution control equipment in response to stringent air pollution regulations. As a result, no plants will be built without acceptable air pollution controls.

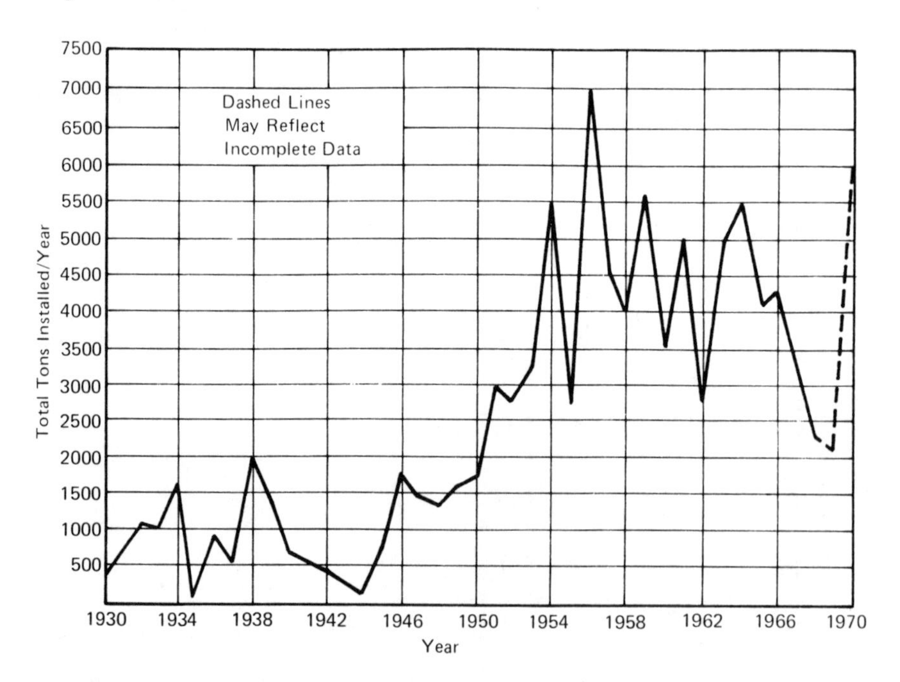

Figure 3-8. Total Annual Additions to U.S. Incinerator Capacity. Source: Niessen *et al.*, *Systems Study . . .*, p. I-26.

Table 3-2
Incinerators Built Between 1940 and 1970 in the United States, by Stoker Type

Number Built	*Feed and Stoker Type*
97	Batch–Circular
82	Batch–Rectangular
49	Batch–Hearth
11	Continuous–Grate-Kiln
14	Continuous–Reciprocating Grate
39	Continuous–Traveling Grate
24	Continuous–Rocking Grate
316	

Source: Niessen *et. al., Systems Study* . . .

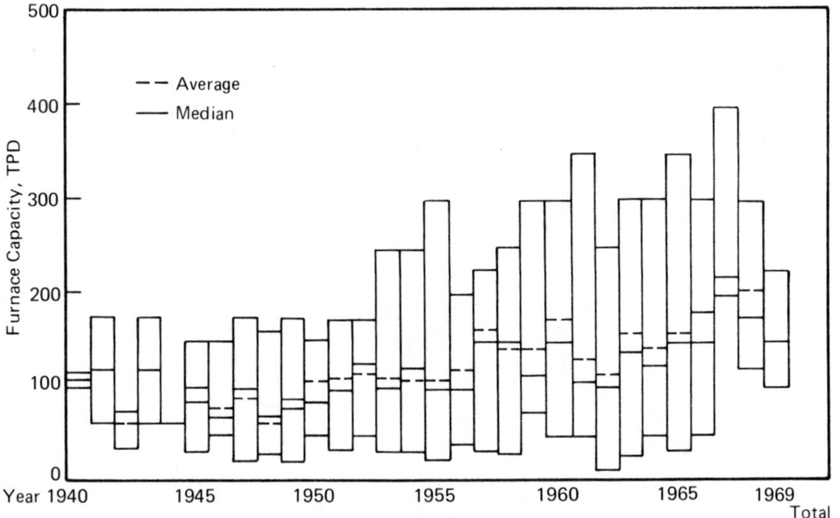

Figure 3-9. Range of Furnace Capacities: New, Rebuilt, Additions to Existing Plants. Source: Niessen *et al., Systems Study* . . ., p. I-29.

In addition to air pollution, it is now becoming recognized that incinerators are also contributors to water pollution. Of the 205 replies to the 1966 survey, only 72 responded to the question on treatment of waste water (Table 3-3).

In terms of refuse storage and handling prior to incineration, the 1966 survey also collected the storage capacity data in terms of operating hours

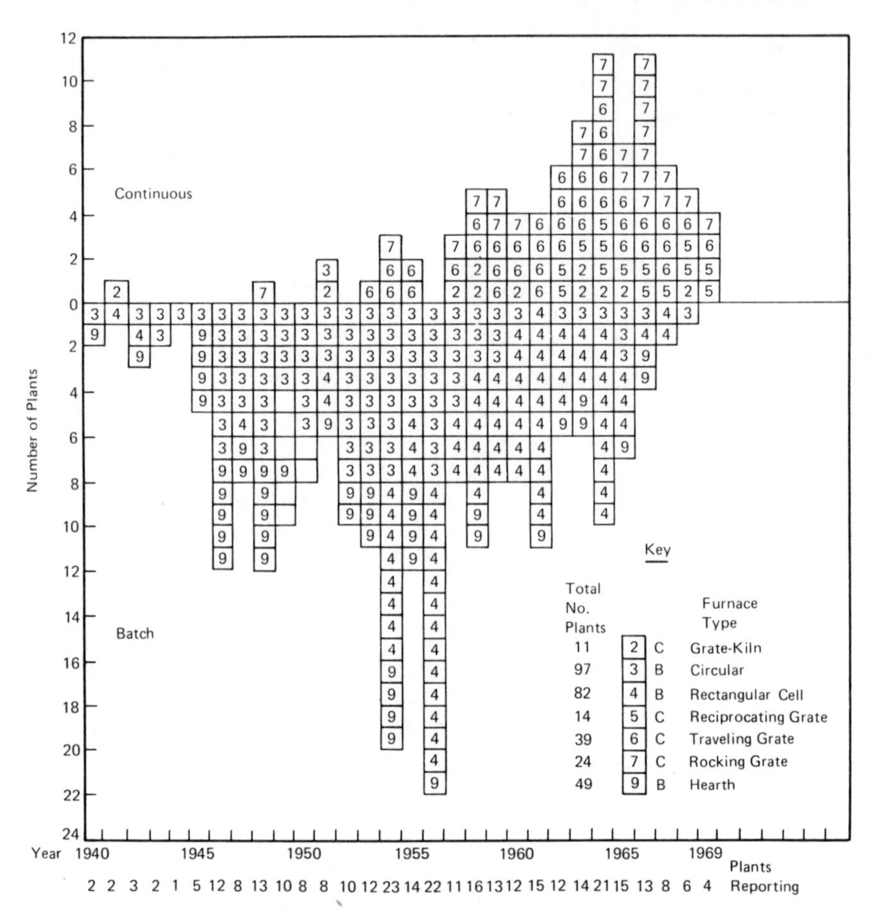

Figure 3-10. Stoker Type and Furnace Feed. Source: Niessen *et al., Systems Study . . .*, p. I-33.

for the reporting incinerators. As can be seen from Table 3-4, the trend is to build incinerators with larger storage capacity. Table 3-5 shows how the incinerators are receiving their refuse. The overwhelming trend is to use the bin and crane. Table 3-6 shows the number of cranes versus the plant capacity for the reporting incinerators.

In terms of waste heat utilization, the 1970 survey indicated that there were only 33 incinerators using waste heat in any form. The 1966 survey, on the other hand, indicated that 43 incinerators utilized waste heat. The apparent discrepancy may merely be one of semantics regarding the word "use." In any case, Table 3-7 shows that the majority of those 43 incinerators did not use the heat for outside-plant purposes.

Table 3-3
Waste Water Treatment and Disposal

	Plants Reporting			Treatment[a]				Disposal							
	Total	*Treatment*[a]	*Disposal Untreated*	*Settling*	*Lagoon*	*Centrifugal Separator*	*Other*	*Sewer*	*Water-course*[c]	*Re-circulated*	*Run-off*	*Sewer*	*WPCP*[d]	*Re-circulated*	*Wasted*
1945–1950	1	0	0												
1951–1955	11	4	2	4				2		1	1	1		1	
1956–1960	22	14	2	10	3	1		4	1	3				1	1
1961 to Date	38	27	10	13	6	5	3[b]	1	3	12	—	3	1	4	2
Totals	72	45	14	27	9	6	3	7	4	16	1	4	1	6	3

[a] At incinerator plant.
[b] Strainers, 1 plant. Settling, screening, and filtering prior to recirculation, 2 plants.
[c] River, stream, bay, etc.
[d] Water pollution control plant.
Source: Stephenson and Cafiero, "Municipal Incinerator Design Practices and Trends," *Proceedings of the National Incinerator Conference* (1966), ASME, p. 15.

Table 3–4
Refuse Storage Capacity (Hours' Operation at Design Rate)

	8-Hour Operation				*9 to 16-Hour Operation*				*17 to 24-Hour Operation*			
	No. of Plants	*Max.*	*Min.*	*Median*	*No. of Plants*	*Max.*	*Min.*	*Median*	*No. of Plants*	*Max.*	*Min.*	*Median*
1945–1955	7	24.0	10.5	12.0	2	16	11.3	–	9	30	11.5	20.5
1956–1960	6	22.5	6.5	10.0	3	25	14.0	16	16	44	8.0	24.0
1961 to Date	5	27.0	15.0	17.5	2	32	16.0	–	29	77	18.0	29.0

Source: Stephenson and Cafiero, "Municipal Incinerator Design . . .," p. 5.

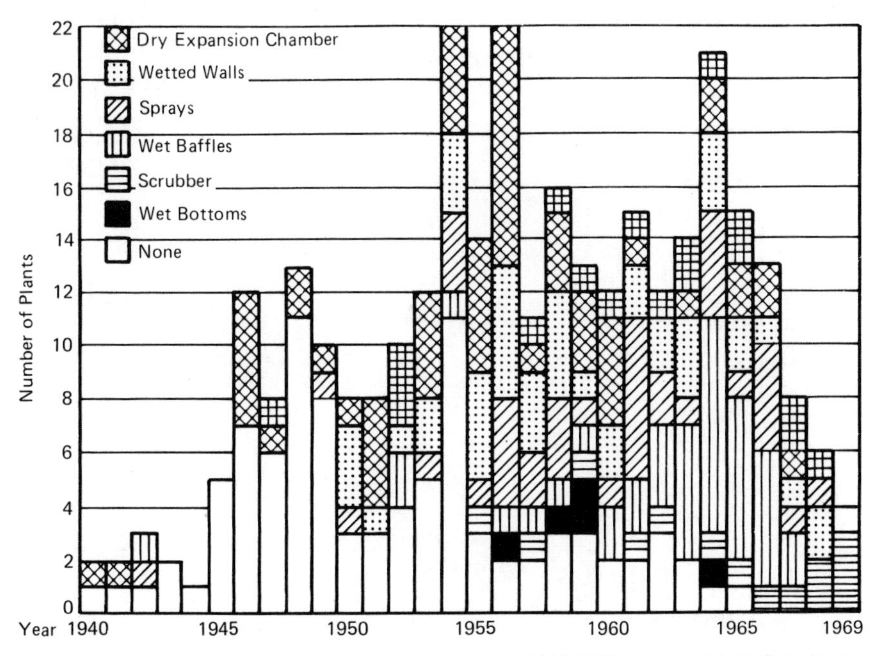

Figure 3-11. Primary Air Pollution Control Equipment. Source: Niessen *et al., Systems Study* . . . , p. I-34.

Table 3-5
Methods of Receiving Refuse

	Floor Dump		*Bin and Crane*	
	Number	*Percent*	*Number*	*Percent*
1945-1955	28	42.5	38	57.5
1956-1960	5	10.0	44	90.0
1961 to Date	4	7.0	56	93.0
Total	37		138	

Source: Stephenson and Cafiero, "Municipal Incinerator Design . . . ," p. 4.

Table 3-6
Number of Cranes Versus Plant Capacity

Plant Capacity T/24 Hr.	*One Monorail Hoist*	*One Bridge Crane*	*Two Bridge Cranes*	*Three Bridge Cranes*
90	1	1		
100–180	5	14		
200	2	11		
225–250		11		
300		17		
320–350		2		
360		3	1	
375–390		2		
400–480		8	8	
500		1	12	2
600–700			10	
720–900			5	4
1000			4	
1050				1
1200			1	2
Total	8	70	41	9

Source: Stephenson and Cafiero, "Municipal Incinerator Design . . . ," p. 5.

Table 3–7
Waste Heat Utilization

	Plants Reporting Use	Number of Plants Using Heat For:					
		Steam Production					
		For Sale	*Outside Heating*	*Other Use*	*Use Not Stated*	*Preheat Combustion Air*	*Other Use*
1945–1950	2	0	0	0	0	0	0
1951–1955	10[a]	1	0	0	2	1	0
1956–1960	17[b]	0	1	[d,g]	1	1	[e,f]
1961 to Date	14[c]	1	1	[h]	1	0	[i]
Totals	43	2	2	3	4	2	3

[a]One plant reports building heat, hot water, and preheating combustion air. Another plant reports building heat and sludge drying.

[b]One plant each reports hot water and power generation; hot water and air preheating; hot water and sludge drying; and steam for equipment drives and heating nearby hospital.

[c]One plant reports building heat and steam for sale. One reports power generation and desalination.

[d]Equipment drives.

[e]Sludge furnace.

[f]Heat for sludge digester.

[g]For sewage treatment plant.

[h]Desalination of sea water.

[i]Tubular gas reheater cools combustion-chamber outlet gas and reheats scrubber exit gas.

Source: Stephenson and Cafiero, "Municipal Incinerator Design . . . ," p. 15.

4 Refuse Selection and Preparation Before Burning

Every mechanical process has some limitations placed on it—especially by the raw materials which it can handle. Incinerators are no exception. Rules are required that prevent non-combustible construction waste, such as concrete, bricks, pipe, and steel beams, from entering the furnaces. Household refrigerators, ranges, bed springs, mattresses, washing machines, and dryers are also taboo. Large and heavy automobile parts, truck tires, drums of oil and tar, and 55-gallon steel drums are prohibited, primarily because they are too bulky for the grates and ash handling equipment.

Conventional municipal incinerators are also not designed to burn truckloads of machine shop scrap, wet grass, tree branches, logs, stumps and heavy timbers, or wet sludge. For example, an incinerator in Delaware County, Pennsylvania, was severely damaged when a load of magnesium turnings from a local machine shop was loaded onto the incinerator grate. Such materials should normally be disposed of by landfill.

While most of the refuse is transferred directly from the pit to the furnace-charging hoppers, the crane is also used to mix the drier commercial/industrial waste with the more moist household waste by rehandling the material in the pit before lifting it to the hoppers.[1] When the industrial waste is quite dry, some plants even spray water on the refuse at this point to aid in holding down peak furnace temperatures.

Despite rules and precautions, some oversized material escapes the attention of the crane operator and passes through the four-foot wide chute and is later discharged onto the grate. While some incinerators can automatically discharge a 55-gallon oil drum or other objects within the same size limits, it is usually necessary to remove the oversized object through a door near the discharge end of the grate.

Refuse Shredders

For a number of reasons then, the practice of mechanically reducing the size of refuse before incineration is receiving a great deal of attention. In general, this process is known by a number of names, including impacting, shearing, tearing, grinding, milling, shredding, pulverizing, and flailing. Shredding, however, is the most commonly used term and generally refers to any mechanical unit that accomplishes the dry liberation, size reduction, and homogenization of solid waste. The municipal solid waste shredder normally uses a combination of techniques including impacting, tearing, and shearing. Various manufacturers rely on different forms of reduction. For example, the Eidal shredders mainly use a grinding action, whereas the Heil/Tollemache shredders employ an impacting action. The results are basically the same: both machines are capable of reducing municipal solid waste to a nominal size of approximately three inches.

The two most popular types of machines used for shredding municipal solid waste are *hammermills* and *ring grinders* (shown in Figure 4–1). The term "hammermill" describes any machine in which a number of flailing hammers are used to strike the material as it is falling through the machine or as the material rests on a stationary metal surface. The struck particles are then thrown against fixed surfaces or are pinched between the moving hammers and the outlet grate bar system. Hammermills employ either a horizontal or vertical rotor shaft. In the more widely used horizontal shaft hammermill, refuse is fed into a large hopper. The primary reduction occurs when the hammers strike the material as it enters the rotor. Further reduction takes place as the material is pinched between the hammers and the grate bars. Oversize material remains in the hammermill until reduced to a size that will permit it to pass through the grate bar openings.

At present, there is very little technical information available on shredding equipment used for municipal solid waste. The facility at Madison, Wisconsin, has provided most of the available data, which is essentially an evaluation of milled refuse in a sanitary landfill, rather than in an incinerator operation.

New data are now being collected at sites throughout the United States on costs and operational problems of shredder facilities. Sites that have accumulated some operational experience include: Columbus, Indiana; Milford, Connecticut; New Castle County, Delaware; St. Louis, Missouri; Vancouver, Washington; Providence, Rhode Island; and Pompano Beach, Florida. Other sites that are now being constructed or are already in operation are at Alamosa, Colorado; Alamogordo, New Mexico; Albuquerque, New Mexico; DeKalb County, Georgia; Guilford County, North Carolina; Onondaga County, New York; San Jose, California; and Willoughby, Ohio. In addition, new sites are planned for Charleston, South Carolina; Ft. Lauderdale, Florida; Garden City, Kansas; Great Falls, Montana; Houston, Texas; Missoula, Montana; Natick, Massachusetts; and San Diego, California.

Many facilities have experienced problems in providing an even flow of refuse to the shredders. It is important to keep an even and optimal feed rate so that the shredder can operate at peak capacity. At some installations, this is accomplished through monitoring systems that feed back information on overload conditions. Most of these systems operate on the amount of electrical current drawn by the shredder motor. If the current is close to the full-load

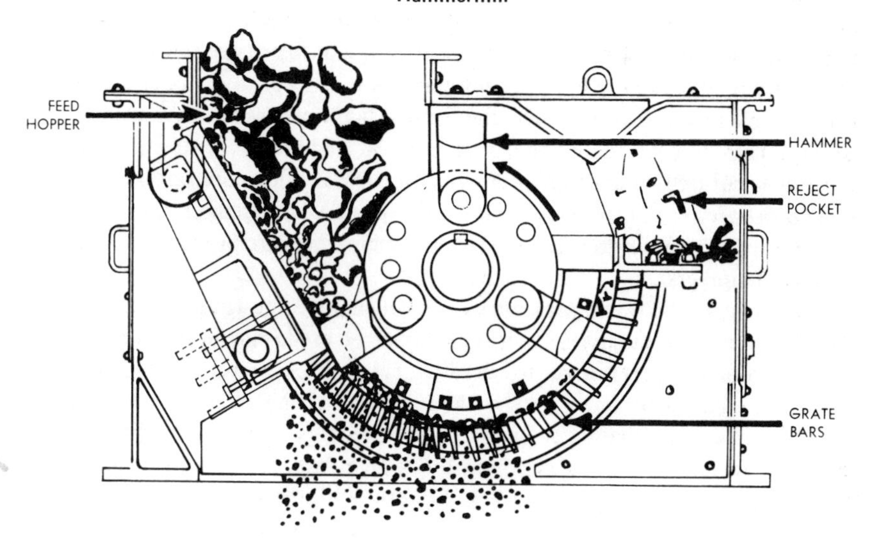

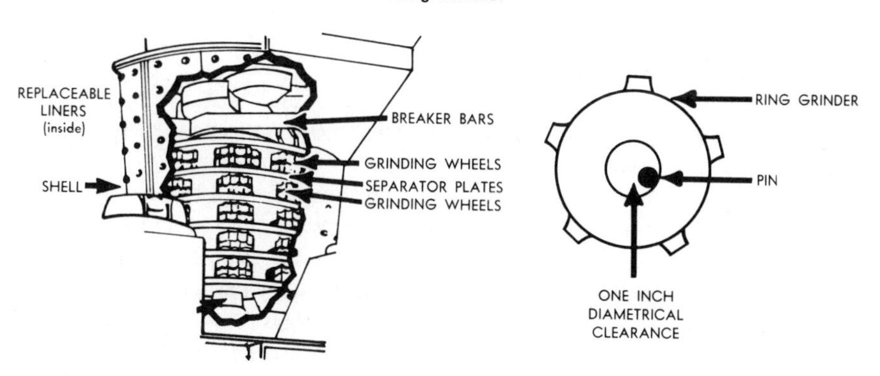

Figure 4-1. Two Types of Machines for Shredding Solid Municipal Waste.

rating, the conveyor is automatically slowed down or stopped, which causes less refuse to be fed into the shredder and prevents overload. Another approach is to use a scale that weighs the amount of refuse on the conveyor belt. If a portion of the refuse has an exceptionally high density, the conveyor is automatically turned off. This type of device is installed at the St. Louis, Missouri, facility.

Most facilities try to remove undesirable materials by hand picking potentially hazardous objects before they enter the shredder. But, because visual inspection is unable to remove 100 percent of the problem materials, some shredders are equipped with reject devices. The most typical is a pocket into which heavy materials are thrown due to their high velocity after impacting with the hammers. To reduce the amount of rejected material, the pocket is sometimes covered with a trap door, which increases the amount of inertia needed to enter the reject pocket. Another reject device, used on the Gondard Mill, is a reject chute into which unusually hard objects are propelled by impact from the hammers.

Municipal solid waste presents significant shredder design problems. In most industries, shredder selection is based on the predominant characteristics of the feed material; is it hard, tough, ductile, brittle, soft, wet or fibrous? Municipal solid waste can, and often does, have all of these characteristics. Some of the specifications that exist for operating shredders in the United States are shown in Table 4–1. Note the extreme differences in specifications, such as rotor speeds that vary from 369 to 1,350 rpm. Within each of these units, there are also a great number of adjustable variables that affect the performance of any unit.

Incinerating Shredded Waste

A municipality's interest in shredding will depend on its basic plan for solid waste management. In the past, the major use of shredders by communities has been for the preparation of compost, although composting has not enjoyed much commercial success in the United States. Because of limited markets, few such facilities are still operating. While not all composting plants employ shredding, the process is normally used in order to reduce particle size to facilitate handling, digestion, and mixing of the materials.

Recently, incinerators have been designed specifically for burning both industrial and municipal shredded refuse. Although the burning of shredded bark and bagasse has been practiced for years, it has only been in the last two years that all-shredded-refuse-fired incinerators have become operational. One industrial facility is being operated by Eastman Kodak in Rochester, New York, and one municipal facility by the City of Hamilton, Ontario, Canada. Another all-shredded-refuse industrial incinerator at the General Motors plant in Pontiac, Michigan, is scheduled to become operational in 1974, and a large municipal plant is scheduled to be completed in Saugus, Massachusetts, in early 1975.

The Hamilton plant is an illustration of techniques that can be implemented when an incinerator is designed for shredded refuse. For example:

1. No overhead cranes are necessary as all of the refuse handling is done by automatic conveyor belts.

Table 4-1
Specifications of Selected Solid Municipal Waste Shredders

	Tons Per Hour	*Horse Power*	*Rotor Speed*	*Discharge Opening*	*Hammer Weight (Lbs.)*	*No. of Hammers*	*Nominal Size of Output (Inches)*	*No. of Shredder Units*
St. Louis	35–75	1250	900	2¼ × 3¼ in.	200	30	1½	1
Houston	40–50	500	900	6½ × 11 in.	100		3	1
San Diego	20–25	500	850	box-type	96	22	5	1
Madison	8–17		1250	155 mm.	16		3	1
	15–25	200	1350	10 × 18 in.	14		3	1
Pompano	22	200	1350		14		3	1
Wilmington	100	900	900	6 × 72 in.	120	33	4	4
Milford	60–80	2 (500)	369		64*	60*	3	2
Vancouver	40–50	1000	369		64*	60*	3	1

*Ring Grinders

2. Shredded refuse can be burned with less air and therefore the gas passage—boiler bank, heat trap, flue and cleanup system—can be made smaller.
3. The shredded refuse allows for a dry ash handling system. There is, therefore, no need for water treatment and drying equipment as the ash can be removed pneumatically, with the air flow providing the necessary cooling element.
4. Better burnout is achieved than with conventional mass burning incinerators.

Because the Hamilton plant is new, actual cost data are not available. However, preliminary estimates indicate that the operating costs for the complete facility are slightly less than those for conventional incinerators.

Burning Shredded Refuse

In theory, there are a number of significant benefits to burning shredded refuse rather than unshredded refuse; these benefits include better surface-area-to-volume ratios, simpler ash handling equipment, and elimination of hot spots through better refuse mixing. Further, shredding refuse represents the first step in any subsequent resource recovery process.

Little is really known about the problems of deep-pile burning shredded refuse. Suspension firing of shredded refuse in existing deep-pile burning incinerators is impractical, as it requires major incinerator modifications; this would, in most cases, essentially result in the rebuilding of the incinerator. While suspension firing shredded refuse in existing incinerators is not practical, it appears that it might be possible to burn the shredded refuse on the grates with only minor modifications. (Only newly designed or recently built incinerators, such as the one in Hamilton, Ontario, just described, are specifically planned to handle all shredded refuse. The Hamilton facility, built in recognition of the inherent advantages of using a shredded refuse feedstock, uses a spreader stoker.) However, there does appear to be a great diversity of opinion among people knowledgeable about incinerators over whether a feedstock of shredded refuse will enhance incinerator performance or whether it will even burn on the grates.

A number of innovative incinerator operators desiring to improve performance have briefly experimented with burning shredded refuse. The cities we have identified as having shredded refuse experiments are Toronto, Canada; Louisville, Kentucky; Rocky River, Ohio; Ft. Lauderdale, Florida; and New York, New York. None of these cities had published the results of their experiments as of late 1973.

5 Salvage Values from Incinerator Residue

Since the Resource Recovery Act of 1970, the national policy has been to encourage separation and salvage of as many components of solid waste as possible. Although a detailed description of separation techniques and salvage markets is beyond the scope of this study, those processes now being designed for future incineration plants will be reported.

As previously mentioned, incineration is not an absolute method of solid waste disposal; even the most efficient incineration leaves a residue. In the past, the primary means of dealing with this residue has been by burial in a dump or landfill. This is still widely practiced and will probably continue for the near future. Using incinerator residue for landfill appears to have much to recommend it, especially when compared with landfilling unburned refuse. The residue from a good incinerator is, for all practical purposes, free of putrescible materials. It is a relatively sterile, inert, and stable fill material. In addition, of course, its volume is substantially less than that of unprocessed refuse; thus, not only are the land requirements for disposal considerably lessened but also the transportation required to bring the material to the fill area, if it is remotely located.

Mineral Resource

Recently, much interest has centered on the recovery of salvageable materials from incinerator residue. The Bureau of Mines of the U.S. Department of the Interior has been active in this field. The need for this kind of effort and some background information has been summarized in the Bureau of Mines Report, *Composition and Characteristics of Municipal Incinerator Residue,* by C.B. Kenahan *et al.,* as follows:

> Residues should not be considered worthless solid waste. They constitute a valuable mineral resource. They are richer in metal and mineral value than many ores. The problem is that the metals in residues are mixed physically and chemically in combinations not normally encountered by extractive metallurgists. Research is urgently needed to develop methods for recovering and refining these metals to prevent further waste of a valuable, readily available raw material. . . .

Information on the composition and nature of incincerated refuse would be useful in determining which type of furnace does the most efficient job of burning. More important, the rapid increase in the generation of municipal wastes, the steady exhaustion of available disposal land within practical hauling distances from population centers, and the high cost of such land have generated a trend toward incineration as a major method of waste disposal, particularly in large urban communities. The trend will intensify as refuse becomes more combustible; at present even poor incineration reduces the volume of the refuse by 75 percent, while efficient burning can reduce it by as much as 95 percent. Salvage of the metallic values from this incinerated refuse would have the further advantage of reducing its volume by another 30–40 percent, reducing hauling costs and greatly increasing the life expectancy of residue landfill sites.[1]

The Bureau has conducted extensive tests on samples of residue from municipal incinerators; their report also makes the following major conclusions:

1. Techniques used in this work demonstrate that sampling of municipal incinerator residues can be accomplished on a relatively small scale with consistent results. The data show that, over a period of several hours, incinerator furnaces will discharge residues of a fairly uniform composition.
2. Glass constitutes the major fraction in all of the samples and accounts for nearly one-half of the residues by weight.
3. Total metallics constitute the second largest fraction and average over 30 percent by weight.
4. Relatively large amounts of unburned paper in some samples, as much as 12 percent, point up the need for more efficient burning.
5. Recovery for recycling of the metallic values could provide significant sources of revenue for municipalities.
6. From a standpoint of reclaiming mineral values, the fine ash fraction is the most promising.
7. Utilization of the metallic and glass fractions, which make up nearly three-fourths of the residues by weight, would reduce greatly the volume of landfill required for disposal. This would extend the life expectancy of residue landfill sites significantly and reduce hauling costs.
8. Recovery and resue of the values contained in municipal incinerator residues would contribute greatly to pollution abatement and aid in the conservation of our dwindling mineral resources.[2]

The Bureau of Mines research to extract resources from incinerator residue has been moderately successful.[3] It has resulted in a continuous process of screening, grinding, magnetic separation, shredding, gravity separation, and other purification steps to recover iron concentrates, aluminum copper-zinc mixtures, clear and colored glass fractions and carbonaceous ash tailings (shown in Figure 5-1). Table 5-1 indicates the composition of incinerator residue on a dry basis.

Although the system is undergoing constant modifications, initial size classification of the incinerator residue is currently accomplished with a trommel and several screens. Ferrous materials are removed magnetically. Large non-magnetic materials are screened. The non-ferrous fraction, primarily glass, is further ground in a rodmill with the glass, being friable, reduced to the con-

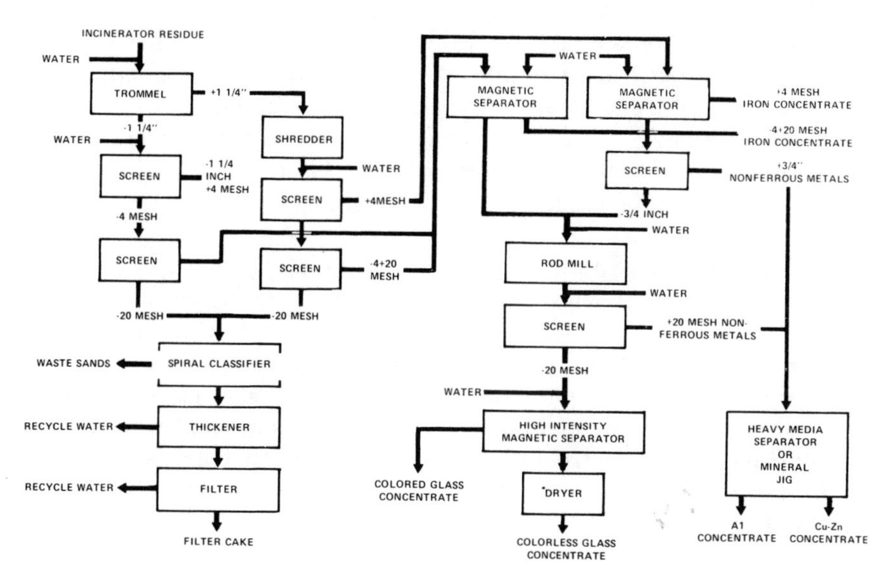

Figure 5-1. Incinerator Residue Processing Flow Sheet.

Table 5-1
Analysis of the Incinerator Residue, Dry Basis

Residue	*Percent*
Wire and Large Iron	3.0
Tin Cans	13.6
Small Ferrous Metal	13.9
Non-ferrous Metal	2.8
Glass	49.6
Ash	17.1
Total	100.0

Source: John J. Henn and Frank A. Peters, *Cost Evaluation of a Metal and Mineral Recovery Process for Treating Municipal Incinerator Residues,* p. 3.

sistency of sand. After screening out the non-ferrous, the glass is separated into colored and colorless elements by use of a high-intensity magnet. The non-ferrous metals are separated either using a heavy-media separator, or a jig, either of which extracts the aluminum from the other non-ferrous metals. Table 5-2 indicates the Bureau of Mines expectations for the value of the recovered products. These values compare with an estimated capital cost of $1.9 million and operating costs of $13.30 per ton for a 400-ton-per-day plant operating on a single shift.

As mentioned previously, the City of Lowell, Massachusetts, in conjunction with Raytheon Corporation, was awarded an EPA grant in September, 1972, to construct a plant based on the Bureau of Mines system to provide 250 tons of incinerator residue per day—the equivalent of 2200 tons of municipal solid waste. Recovered materials will include ferrous and non-ferrous metals and glass. A detailed discussion of these markets is beyond the scope of this report on incineration, but a companion report by the National Center for Resource Recovery, *Resource Recovery from Municipal Solid Waste,* [4] deals exclusively with this issue: the separation, salvage, and recycling of solid waste components.

There appear to be certain advantages to performing resource recovery prior to incineration. During incineration, the tin from the tin cans alloys with the ferrous metal and eliminates the possibility of detinning the recovered material. Elements such as copper also fuse with the ferrous and further reduce potential markets for this material. Some of the aluminum is oxidized during incineration; and the glass recovered from the incinerator residue is dirty and difficult to clean. And, of course, it is impossible to recover paper after incineration. For these reasons, resource recovery before incineration—either on shredded or natural refuse—appears more attractive.

Table 5-2
Estimated Product Values

Product	*$ Value*	*Quantity, lb/ton Residue*	*$ Value/ton Residue*
Ferrous Metal	10/ton	610	3.05
Aluminum	0.12/lb	32	3.84
Copper-Zinc	0.19/lb	24	4.56
Colorless Glass	12/ton	552	3.31
Colored Glass	5/ton	398	1.00
Total Value			15.76

Source: John J. Henn and Frank A. Peters, *Cost Evaluation of a Metal and Mineral Recovery Process for Treating Municipal Incinerator Residues,* p. 12.

6 Representative Incinerator Operations

This chapter presents a few of the many modern instances of incinerator installations. Although many more might have been presented, the handful cited here have each been chosen because they represent a unique facet or advantage of incinerator technology. Thus, Issy-les-Moulineaux, near Paris, has been chosen because this operation is one of the first successful applications of heat recovery. The Norfolk installation was the first U.S. water-wall incinerator. The Nashville operation has been presented because of its unique financing arrangement.

Usine D'Ivry

This plant is the newest of three incineration plants serving Paris and its suburbs and is probably the largest and most modern installation of its type in the world. Each boiler can burn 1200 metric tons (1 metric ton = 2200 pounds) of unsorted, unsized refuse per day (based on garbage with a LHV equal to 4400 BTUs per pound) and produce 278,000 pounds per hour of 1380 psig, 885 degrees F. steam. The steam from the plant can produce 64MW of electricity when operating straight condensing or up to 90 percent of the steam can be extracted from the cycle at 300 psig to help heat the enormous complex of buildings in Paris. The plant was designed to incinerate 600,000 metric tons of refuse per year. A television camera continually scans the rear of the stoker to show the control room operators the condition of the fires. It is reported that only 2 or 3 percent combustible is found in the bottom ash. Management reports that some corrosion problems have occurred on the water-wall furnace. To eliminate this problem, they have studded and refractory-covered the furnace up to the bottom of the furnace screen. The maximum gas temperature leaving the furnace is about 1800 degrees F., and the temperature leaving the economizer is about 570 degrees F.

Magnetic separators are used to retrieve the ferrous fraction, which is sold to nearby steel companies. The rest of the bottom ash is sold to stabilize road beds in highway construction. The fly ash from the electrostatic precipitator is sent to landfills.

Issy-les-Moulineaux

This installation has been in operation for over six years, with the heat produced being utilized to generate power. It was one of the world's first operational applications of heat recovery techniques.

The operation of the plant begins with the unloading of refuse from collection vehicles into a 7,850-cubic-yard pit. There is no provision for sorting refuse at this point. The furnaces are capable of burning everything hauled by the collection vehicles. After collection, the pit is sealed and a negative atmosphere pressure is maintained; all the combustion air is drawn from the pit. Consequently, no dust spreads to the outside. From the pit, the refuse is fed into furnace charging chutes by grab buckets with 6.5-cubic-yard capacity from two traveling cranes. The four furnaces are contained in a part of the building that is 308 × 197 feet. Each furnace is designed to burn 18 metric tons per hour.[1] Each also contains a stoker with a surface area of 570 square feet, divided into three sections, each of which contains 15 steps. These stokers provide continuous tumbling of the refuse through the alternating movement of stationary and mobile bars. The furnaces are formed entirely by water walls. The upper and lower walls have a length of approximately 29.5 feet and a diameter of about 5 feet. The boiler surface is about 10,700 square feet; the superheaters add about 8500 square feet. Although each furnace has oil burners for start-up, they have never been used. Steam is generated at 925 psig and 770 degrees F. The capacity of each of the four boilers is 88,000 pounds per hour. They discharge into a common header from which all the steam passes into a back-pressure turbine that expands the steam to approximately 285 psig. This turbine drives a 9000-kilowatt generator. Steam leaving the turbine is sold at 285 psig to the district heating company serving the entire center of Paris. If the city cannot absorb all the steam, the surplus is taken to a second turbo-generator—a condensing set of 16,000-kilowatt capacity.

From the furnaces, ashes fall by gravity into a water trough from which a piston-operated pusher discharges them onto a belt conveyor. After incineration, the residue is taken to a 2000-cubic-yard storage area where it is generally sold for road construction. The unburned carbon in the residue, including fly ash, is measured to be 3 percent, of which putrescible matter is about 0.1 percent.

Combustion air from the refuse pit is brought into the combustion chamber by compartments under each stoker. The gas temperature at the top of the primary combustion chamber is 1700 degrees F. On leaving the boilers, the combustion products pass through electrostatic precipitators of two fields, which have an efficiency of more than 98 percent. As a result, smoke from the two 260-foot chimney stacks is hardly ever visible. The total dust emission from the incinerator plant is around 187 pounds per hour.

In 1967, the installation had disposed of more than 500,000 metric tons

(564,000 U.S. tons) of a lower heating value of approximately 1800 kcal/kg (3240 BTU per pound). Of this amount, the plant incinerated 556,600 tons, with the remainder sent to a sanitary landfill. During 1967, the plant sold 1100 million pounds of steam and generated 75 million kilowatt-hours of electricity.

Approximately 67 percent of the operating costs are recovered via the sale of steam. The management has found that when plants are on a large scale, the revenue from the sale of steam or electricity is higher than the investment charges, maintenance, and operating costs of the additional equipment required to recover waste heat.

Norfolk Naval Base

The Norfolk Naval Base incinerator at Norfolk, Virginia, is distinctive because it is the first North American incinerator with a water-wall furnace; most water-wall furnaces are found in Europe. Table 6-1 is a complete listing of water-wall incinerators in North America. One major advantage of the water-wall furnace over conventional refractory furnaces is its increased heat transfer efficiency. The design enables temperatures to be held within limits that would be exceeded in a refractory furnace without large quantities of air. Most American refractory-lined incinerators require as much as 200 percent excess air, whereas water-wall furnaces generally operate most satisfactorily at approximately 40 to 80 percent excess air. This type of operation considerably reduces air pollution.

Input into each of the two Norfolk incinerator boilers is 180 tons per day of refuse with an output of 50,000 pounds of steam per hour per boiler. The capital cost was $2.135 million, of which $1.0 million went into equipment. The system has been in operation since 1967 and is estimated to save $47,000 a year in steam production, which represents a savings of over 5 million gallons of fuel a year.

It should be noted that the composition of municipal waste entering the incinerator consists of a large quantity of crates and packaging materials, in addition to normal rubbish, and has an average water content of 25 percent. Although a continuous supply of refuse is essential to the Norfolk operation, the system also includes an auxiliary oil burner that can produce 60,000 pounds of steam per hour at 275 psig.[2]

Chicago Northwest Incinerator

Chicago's Northwest Incinerator is the largest of its kind in the United States and was completed in 1971 at a cost of $23 million. The plant is highly

Table 6–1
Steam-Producing Municipal Incinerators in North America

Name	*No. of Units*	*Date of Service*	*Boiler*	*Stoker Manufacturer*	*Type of Stoker*
City of Montreal, Montreal, Canada Des Carrieres Plant	4	end of 1970	Design by Von Roll–supplied by Dominion Bridge	Von Roll	Reciprocating Grate
City of Chicago, Chicago, Illinois (Northwest Incinerator)	4	end of 1970	Design by Walther Co.–supplied by IBW	Martin	Reciprocating Grate
City of Harrisburg, Harrisburg, Pennsylvania	2		IBW	Martin	Reciprocating Grate
City of Hamilton, Ontario, Canada	2	Summer 1972	B&W Canada	Detroit	Traveling Grate Spreader Stoker
Norfolk Navy Yard, Norfolk, Virginia	2	1967	Foster–Wheeler	Detroit	Reciprocating Grate
Braintree, Mass.	2	Spring 1971	Riley		Traveling Grate
Quebec City, Canada			Design by Von Roll–supplied by Dominion Bridge	Von Roll	Reciprocating Grate
City of Nashville, Nashville, Tennessee	2	1973	B&W	Detroit	Reciprocating Grate

automated and designed to handle 1600 tons of refuse per day, operating 24 hours per day, seven days a week. The facility is constructed on an 11-acre site near a residential area and is architecturally designed to contain the offensive odors from the raw refuse. Further, its double-wall construction reduces the noise of the incineration and shredding processes in the surrounding area. The incinerator is a modern water-wall construction.

Steam is generated in four burners within the incinerator, each capable of generating approximately 110,000 pounds of steam per hour. All refuse, with the exception of large appliances and bulky items such as mattresses, is burned in the "as received" condition. The larger items are processed through a turbine-driven, 25-ton-per-hour bulk refuse shredder, which reduces these bulky items to pieces 6 inches in size. A magnetic separator removes the shredded ferrous

Tons of Refuse Per Day Per Unit	*Steam Flow Per Unit lb/hr.*	*Steam Pres. Psig.*	*Total Steam Temp. °F*	*Remarks*
300	100,000	225	500	Burnout reported as better than 99.5 percent and emission from elec. precip. below 0.2 pounds per 1000 pounds of gas.
400	110,000	275	Sat.	Metcalf & Eddy Construction Engineers, Boston, Mass.
360	98,000			Garnett, Fleming, Corddry and Carpenter, Inc., Harrisburg, Pennsylvania
300	106,000	275	Sat.	Gordon L. Suton & Assoc., Ltd.
180	50,000			Metcalf & Eddy Construction Engineers.
120			Sat.	
				Duplicate of City of Montreal.
360	109,000	400	600	I.C. Thomasson Assoc., Engineers.

metal and the remainder of the refuse is combined with the ordinary refuse in a discharge pit. Three cranes equipped with 5-cubic-yard buckets pick up the refuse and drop it into the furnace-feed hopper. Once in the hopper, the burning process becomes automatic. The refuse is fed onto the stoker grates by means of a hydraulically fed ramp, controlled by the pressure in the under-grate air plenums. Stoker grates are of the reverse reciprocating type inclined at a 26-degree angle that produces a downward flow of refuse. At the same time, the reverse-acting grate bars push the refuse back up the incline to create a tumbling and mixing action for maximum combustion of the refuse. The water-filled ash discharger is located below the stoker grate. The water both quenches the ashes and seals the discharger against the escape of gases. The hot gases then pass through the boilers and on to electrostatic precipitators. Of the 440,000 pounds of steam generated by the four boilers, 240,000 pounds are used for in-plant needs, such as operating the turbine-driven machinery. The remaining 200,000 pounds of steam are available for sale.

The temperatures within the boiler reach 1500 to 2000 degrees F. The hot gases, after leaving the boiler, are approximately 450 degrees F. at the entrance to the electrostatic precipitators. It is estimated that 97 percent of the particulate matter contained in the gases is removed by the precipitators. It is significant that, with this incinerator, Chicago became the first major city with the capability of incinerating all its municipal refuse.[3]

Hamilton, Ontario, Solid Waste Reduction Unit

The installation in Hamilton, Ontario, is the only all-shredded municipal refuse-fired incinerator in the world. There, collecting vehicles unload the refuse into a 40- by 80-foot deep pit with a capacity of 450 to 500 tons.[4] The bottom of the 30-foot-deep pit is made up of four 84-inch wide apron conveyors that take the refuse to four Tollemache shredders. Each 200-horsepower shredder is electrically driven and has a capacity of 15 tons per hour. As the refuse moves toward the shredder, two men remove bulky items that are trucked to a landfill. The shredder reduces the refuse to particles of two inches in size. Dust generated by grinding is drawn down by the fan of the rotor and kept in the material flow stream.

From the shredder, the refuse is conveyed past two magnetic drum separators that pull out the ferrous metals to save wear on the incinerators. This ferrous fraction is presently being sold to scrap dealers. From the magnetic separator, the refuse goes on belt conveyors to a 700-ton Atlas storage tank. The storage tank feeds a single belt conveyor that transports the refuse to three feed chutes at the top of the boiler. The refuse enters about 10 feet above a traveling grate. It is blown into the furnace with a blast of air from three feed nozzles mounted in the front of the furnace. It is estimated that 50 to 70 percent of the material is burned before it hits the grate below. Burning in the furnace reduces the shredded refuse to about 10 percent of its original volume. Complete combustion occurs before the ash drops into the ash pit. The ash, which is made of a fine grit with no trace of putrescible matter, is pneumatically removed. Electrostatic precipitators are used for cleaning the combustion gas; there is also a bag house for dust control.

The incinerator plant is designed to dispose of 600 tons per day at full capacity. This amounts to an estimated disposal of 200,000 tons of refuse per year. At full capacity, the furnaces will operate 24 hours a day with the pulverizers working about 10 hours. The plant has two water-wall incinerator boilers, each capable of generating 100,000 pounds of saturated steam per hour at

250 psig. All in-plant equipment larger than 25 horsepower, with the exception of the shredders, is powered by the steam turbines.[5] This equipment uses 10 percent of the steam produced. Although the remainder is available for sale, it is currently condensed.

The Hamilton plant had an original capital cost of $8 million. It is estimated that the capacity could be doubled with an additional investment of $4 million. In that case, the total investment for a 1200-ton-per-day plant would be $12 million. The present operating cost is approximately $3.50 per ton with depreciation and interest amounting to $4 to $4.50 per ton.

Nashville Thermal Transfer Corporation

As a result of Nashville's needs for a central heating and cooling system and a more adequate way to dispose of municipal solid waste, the city established the Nashville Thermal Transfer Corporation. Expected to begin operation by mid-1974, the plant will incinerate solid waste and use the recovered heat to produce steam and chilled water for municipal use.

Plant expansion will take place in two steps. In the first phase, the plant will consume 720 tons of solid waste per day in two water-wall incinerator boilers, producing 215,000 pounds of steam per hour.[6] The steam will be used for heating and driving two 7,000-ton centrifugal refrigerating units with a total of 13,500 tons of cooling capacity.[7] Chilled water at 40 degrees F. and steam at 125 psig will be supplied to city buildings through 2.9 miles of underground piping. By 1978, the plant will be capable of handling 1300 tons of waste per day, which is the amount now generated by the City of Nashville. At this point, the plant will be generating 500,000 pounds of steam per hour in four incinerator boilers and have 31,000 tons of cooling capacity in 5 large water chillers. It is estimated that this amount of heat is equivalent to that generated by 400 tons of coal.[8]

In the first step of plant operation, solid waste will be delivered by the city at no expense to the plant under a 30-year contract. The waste will be dumped into refuse pits and picked up in one-ton "bites" by a crane. The crane operator must attempt to exclude bulky, non-combustible items such as engine blocks and large appliances. The waste will then be continuously fed by moving grates into incinerator boilers. To combat the problem of odor, the furnace will be sealed and operated under a slightly negative pressure, and combustion air will be drawn into the plant through waste storage rooms, thereby burning the odors in the incinerators. The use of excess combustion air will maintain a constant furnace temperature of about 1800 degrees F. despite the uneven burning qualities of solid waste.[9] Steam will be generated in the boilers at 400 psig and 600 degrees F. and piped to a second building that will contain monitoring computers and the turbine-driven equipment such as chillers and pumps.

Non-condensing turbines on the pumps will reduce steam pressure to approximately 150 psig before entering the condensing turbines on the chillers.

Nashville planners hope that the thermal transfer plant will produce less air and water pollution than present heating and cooling plants. Dry cyclone collectors will remove particles about 10 microns in size, which make up 30 to 40 percent of all particulate emissions.[10] The remaining particulates and gaseous pollutants will be treated in three-phase wet scrubbers that will also remove sulfer oxides, hydrogen chlorides, and other acidic components. The bleed water from the scrubbers will go to ash quench tanks where it is lost by evaporation and ash dragout. The bleed from cooling towers, boiler blowdown, and plant clean-up water will be fed into sanitary sewers or treated before return to the river.

The $18-million facility was financed by creating a non-profit corporation that sold revenue-type bonds. After bond retirement, the plant will become city property. Table 6-2 gives the projected 1978 costs for production of coolant and steam. The Nashville Thermal Transfer Corporation estimates that at peak capacity, steam will be produced at a cost of 70 percent less than that of a fossil fuel plant; the chilled water, for 60 percent less.

Hempstead, New York

Another interesting incinerator installation is the one at Hempstead, New York, which results in the production of steam and, ultimately, electricity. The steam is produced in two boilers, each having a capacity of 85,000 pounds of steam per hour, and is used to generate 2500 kilowatts of power to operate the plant. The exhaust steam–i.e., steam not used for electrical power–is used to produce approximately 500,000 gallons of fresh water per day. This water is used within the plant itself for ash removal sprays, boiler makeup, and cooling.

Table 6-2
1978 Projected Production Costs for Coolant and Steam by Nashville Thermal Transfer Corporation

	Coolant $/ton-hour	*Steam $/1000 pounds*
Fixed Charges	0.00865	0.306
Operating Cost	0.00950	0.094
Distributing Cost	0.00575	0.152
Totals	0.02390	0.552

The $6-million plant processes 750 tons of refuse per day. The plant has two 300-ton-per-day refractory furnaces, and one 150-ton-per-day furnace for bulk rubbish. Hot gases from the refractory furnaces generate steam in the boilers. The steam is first used in the turbine generators and then circulated through a closed system (four single-stage submerged-tube evaporators) where it is cooled with salt water from a nearby inlet channel. Heat transferred to the evaporators boils the salt water and converts it to steam. The steam is drawn off and condensed to fresh water, while the concentrated brine is pumped back to the inlet channel.

Other Concepts

A good review of other developments is presented in "European Practice in Refuse Burning," by Georg Stabenow, in the *Proceedings of the National Incinerator Conference* (1964), which discusses four other incinerators. The first is the incinerator at St. Ouen, Paris, a rotary-kiln incinerator like that being used successfully in this country in Tampa, Florida. The incinerator at Munich is also discussed. A cut-away detail of the reverse-action grates used at Munich shows them to be the same as those in the Chicago Northwest Incinerator and those in Harrisburg. An incinerator (discussed in Chapter 3) at Dusseldorf that uses drum grates is also presented. At present, there are no incinerators in the United States that are operating with rotary-drum grates.

Other new municipal incinerator plants are in operation in a number of countries around the world. Reviews of them are to be found in the *Proceedings of the National Incinerator Conference* (various years) and in trade journals in the municipal field, such as *American City* and *Public Works.* Steam-generating incinerators are also described in the magazines *Combustion* and *Power.*

7 Heat Recovery

Obviously, a great deal of heat is generated by incineration. The heating value of raw refuse in the United States is reported to average about 4450 BTU per pound.[1] About 65 percent of this heat can be transferred to steam in a boiler. In many cities where refuse is centrally incinerated, there are also large, coal-fired boiler plants generating electricity. The idea of combining the functions of these two plants–using refuse as a fuel to generate power while at the same time disposing of the refuse–has been a matter of widespread discussion for many years. This concept has become common practice in many European countries but has only recently begun to receive much attention in the United States. Such an operation is not simply a matter of changing fuels from coal to refuse. Furthermore, a city's refuse could produce only a small percentage of the electric power required. It seems likely, however, given the urgent need both for increased power and for refuse disposal, that the operational and economic problems will be solved and that we will see much more recovery of heat from incineration in the years to come.

The upward trend in fuel prices, the shortage of natural gas and other fossil fuels, and the growing unpredictability of fuel imports, encourage fuel economy and the use of all domestic heat sources.

There are a number of ways in which heat may be recovered from incinerator operations. One way, as suggested above, would be power plants that would use refuse alone or with fossil fuel to generate steam, which in turn would drive turbo-generators to produce electricity. Alternately, steam produced by an incinerator might be sold to a nearby generating plant. A third possibility is the sale of steam to local industries that could use it either directly for heating or processing or for running generators for internal electric power requirements. Downtown areas and college campuses are also potential markets since the steam could be used for both heating and cooling. Obviously, the alternative selected for any given location must depend on the needs of industries and utilities in the area, and these should be a prime consideration in the design and location of the plant.

Besides the revenue to be obtained from the sale of steam or electric power, there is another possible economic advantage in using the heat from refuse. This is due primarily to recent requirements for the addition of pollution control devices to incinerators. First, the very hot gases emitted from an incinerator furnace must be cooled considerably before they reach pollution control equipment. Conversion of the heat in the gases to other forms of energy accomplishes this need. At the same time, the volume of gas to be cleaned is substantially reduced, and thus the size and cost of the cleaning equipment may be reduced. As stated in the Public Health Service's *Incinerator Guidelines–1969,* "the trend toward more effective and costly pollution control equipment increases the economic feasibility of heat recovery."[2]

In considering the economic feasibility of utilizing heat to generate power, it is important to keep the question in the proper context. Hansen and Rousseau, in "An Engineering Approach to the Waste Disposal Crisis," have stated it as follows:

> The question is not whether such plants can generate steam or electricity at a price competitive with plants built solely for this purpose, but whether the generation of power can reduce the cost of solid waste disposal after deducting credits for the sale of power. In other words, it is important to investigate whether the revenue from the sale of steam or electricity is higher than the investment charges, the maintenance, and operating costs of the supplementary equipment required to recover the waste heat (boilers, water treatment, turbogenerators, transformer stations).[3]

There are several types of systems that have been used for heat recovery from incineration plants. The Public Health Service lists the following four basic designs:

1. Waste heat boiler systems with tubes located beyond conventionally built refractory combustion chambers;
2. Water tube wall combustion chambers;
3. Combination water tube wall and refractory combustion chambers;
4. Integrally constructed boiler and water tube wall combination.[4]

Refractory incineration uses roughly 200 percent excess air to burn refuse, while water-wall units use approximately (a) 80 percent excess air with mass burning and (b) 40 percent excess air with shredded refuse on a spreader stoker. Because the water-tube wall furnaces require less excess air, the installation of smaller, less expensive air pollution control devices is possible.

A number of criteria for modern boiler/incinerator plant features have been suggested. A few of these are summarized below:

It is recommended that steam pressures not exceed 450 psig, and that steam temperatures be kept below 500 degrees F. A short drying arch should be coupled with a long, low burn-out arch, in which the gases enter a throat where they receive intensive mixing with excess air. Furnace temperatures should be not lower than 1500 degrees F.

Table 7-1
Heat Recovery and Boiler Efficiency

Gross Heating Value as Fired, BTU/Lb.	*5,000*	*6,000*	*6,500*
Specific Steam Flow, Lb. Steam/lb. Refuse	3.3	4.1	4.4
Excess Air at Economizer Outlet, %	88.5	88.0	88.0
Total Boiler and Grate Efficiency, %	66.0	67.0	67.5

Source: G. Stabenow, "Performance and Design Data for Large European Refuse Incinerators with Heat Recovery," *Proceedings of the National Incinerator Conference* (1968), ASME, p. 281.

Although the above temperature would probably prevent any high-temperature corrosion of the pressure parts, the low steam pressures generated would preclude generating electricity from the steam. Babcock and Wilcox therefore recommend a compromise of approximately 600 psig 750 degrees F.[5]

The boiler should be designed specifically for incinerator duty, with widely spaced tubes in line for slow, unobstructed gas flow (15 feet per second). The refuse should receive mild agitation, and the grate openings should be small.

Depending on the type of equipment used, and the heating value of the refuse, steam production can range between one and five pounds per pound of solid waste burned. Estimates from one source are presented in Table 7-1.

Operational Considerations

In June 1971, *Solid Waste Report* stated that "the U.S. lags behind all other advanced technology countries in the utilization of the heat content of solid waste in incineration systems through the generation of steam, hot water, or electrical power and in some types of incinerator design."[6] Until mid-1973, it appeared that the United States had ample supplies of fossil fuels at relatively low costs. Up to that time, it was not deemed economical to recover heat from refuse. Of course, that perspective changed with the energy crisis experienced both here and abroad.

Fuel shortages came earlier in Europe. As a result, the available literature has numerous references to foreign experience with heat recovery. An excellent review of the European combination incineration/power station, along with discussions on how refuse might be used as a fuel, can be found in "Systems Evaluation of Refuse as a Low Sulfur Fuel," by Envirogenics Company (November 1971). Likewise, "Recent Developments in Operating Experience at British Incinerator Plants," by R.H. Watson and J.M. Burnett, presents an overview of British incineration, emphasizing heat recovery.[7] The fact is that British incinerators are not too unlike those in the United States and the prospects for heat recovery encounter the same problems as those found in the United States.

For an earlier (1968) discussion of much the same material, see "Performance and Design Data of Large European Refuse Incinerators with Heat Recovery," by G. Stabenow.[8] It presents the state of the art on steam-generating incinerators in Europe as of 1968. Similarly, H. Hilsheimer in "Experience After 20,000 Operating Hours at the Mannheim Incinerator," has detailed the operational results of a combination incinerator/power plant where refuse-fired boilers are tied into oil-fired boilers.[9] There are no similar facilities in the United States.

Nevertheless, even in the United States, conditions are now changing to favor heat recovery rather than use of more fossil fuels. The calorific value and tonnage of refuse is increasing. (Niessen estimates that the per pound heating value and tonnage generated will increase by 17 percent and 151 percent, respectively, between 1970 and the year 2000.)[10] The price of fossil fuels is rising, fuel shortages are appearing, and the design of incinerator/boiler combinations has improved. Furthermore, the trend to more stringent air pollution control favors the use of boilers for cooling the flue gases prior to gas cleaning.

Probably the most frequently stated problem in heat recovery is that of matching the supply of heat to the demand for heat. In fact, in most cases the economics just do not appear favorable. For example, Day and Zimmermann report on an inquiry made to Potomac Electric Power Company in Washington, D.C. with respect to a new incinerator then in the design stage:

> The Potomac Electric Power Co. operates a steam turbine powered electric generating station adjacent to the plant site. They were contacted to determine if they could use a supply of steam from the incinerator plant at the 225 psig dry saturated conditions, available from an incinerator-boiler. We were advised that they could use 200 psig steam in their older turbines but that superheated steam was preferred. The minimum cost of steam generation at the incinerator plant plus pipeline charges is in excess of the steam generating costs for low pressure steam at the utility power plant.[11]

This is a good illustration of why there have been no major examples in this country of incinerator production of steam for electric utility use until the recent St. Louis tests that are discussed later in this chapter.

One reason for the cost differential is the matter of economy of scale. Efficient thermal electric plants are extremely large, burning upwards of 5000 tons of coal per day or an equivalent in BTU value of oil or gas. Except in rare circumstances (New York City, for example), it would be extremely difficult to reach the required scale of operation with refuse as a primary fuel without escalating the cost of solid waste management by increasing transportation costs. The Chicago Northwest Incinerator at 1600 tons per day is the largest in the country. Yet, a 2000-ton-per-day plant would only produce about 40 megawatts of electricity, while labor costs for a plant of this size would be almost the same as one of a 1000-megawatt facility.

Recently, the state of Connecticut announced a competitive procurement to develop a plan to deal with Connecticut's long-range solid waste disposal problem in an energy-conversion approach. Attached to the basic letter is a memo entitled "Utilization of Solid Waste for the Generation of Electricity" and a pledge of not only cooperation but also "substantial effort" from Northwest Utilities, a consortium of light and power companies serving the state.[12]

There is also significant potential in using incinerator heat in combination with a sewage disposal plant: the hot gases from the incinerator furnaces can be used to flash-dry the sludge from sewage disposal plants; the sludge can then be incinerated if it is not to be processed for fertilizer. Furthermore, noxious and odorous gases from the sewage disposal plant can be burned in the incineration plant furnaces before they are discharged into the atmosphere.

J.W. Regan has suggested that waste heat utilization could be improved if the refuse received extensive processing prior to burning. He points out that in the case of coal, for instance, the fuel is "washed, sized, crushed, and pulverized to 65 to 75 percent less than 200 mesh, pneumatically transported to the furnace, and burned in suspension. These methods of preparation and handling are known and accepted as being necessary. Why then do we think we can do a good job of burning an inferior fuel such as refuse in the raw state?"[13] He goes on to suggest a similar procedure for burning refuse for steam generation. It is stated that refuse could be shredded to a 2- by 2-inch size. This size range would, in fact, be a long way from the pulverized coal used in present-day suspension-fired boilers.[14]

Tests are being conducted by the Union Electric Company in St. Louis, Missouri, 1972 to 1974, in which two boilers of their Meramec plant were modified so that they burn prepared refuse along with coal. Approximately 300 tons of refuse a day are consumed, providing 10 percent of the heat value. These boilers have a steam capacity of 950,000 pounds per hour at 1500 psig, 950 degrees F. main steam and 950 degrees F. reheat, with a generation capacity for each boiler of 125 megawatts. The boilers are designed to be tangentially fired with pulverized coal burners in each corner, supplemented by natural gas. For the experiment, a gas nozzle in each corner of the boilers was replaced with a refuse inlet.

In the electric facility itself, the installation was completely funded by Union Electric (approximately $600,000). About $2 million, of which EPA contributed approximately $1.7 million, was spent on the city's processing facility, storage and transfer equipment at the site and unloading and transfer facility on ground leased by the city at the Meramec plant.

Refuse is dumped on the tipping floor at the city facility, and loaded on a conveyor by a front-end loader. The conveyor then feeds the shredder. From the shredder, a Rader air classifier separates light from heavy fraction; the ferrous materials are then extracted from the heavy fraction. The light fraction is conveyed into a storage bin. Discharge from this area is by conveyor to a stationary packer that loads transfer trailers for the 18-mile trip to the power plant. There it is unloaded into another storage receptacle (city-owned). It is removed from this storage by a pneumatic system and transferred several yards to another storage area (Union Electric-owned).

From the storage bin, the refuse is conveyed to four out-feed chutes through four air-locked pneumatic feeders to the four corners of the boiler furnace.

The St. Louis experiment validates the heat value of refuse. Clearly, the utility has potential fuel savings over its alternative of coal or gas, as well as savings for the city over its disposal alternative of incineration.

In February 1974, Union Electric announced plans to construct a $70-million system to receive a major portion of refuse from St. Louis and adjoining counties.

8 Air Pollution Control

For years, air pollution from many sources, including incineration, was either ignored or tolerated as an unavoidable side effect of an industrialized society. During the past several years, however, laws and regulations have been passed at all levels that set maximum emission standards for industrial plants, including municipal incinerators. These laws require the installation of pollution control equipment on practically all plants built after 1965. Plants are now being designed with the pollution control system as an integral part. Obviously, this course is a more economical one than adding on pollution-control equipment after the fact.

The seriousness of the incinerator air pollution problem is illustrated by Table 8-1, which estimates the 1968 incinerator stack emissions along with an estimate of what the typical incinerator emits per ton of refuse incinerated. Note that the emissions from the stack are almost as high as those leaving the furnace.

L.P. Duncan of the MITRE Corporation has completed a nationwide summary of anti-pollution regulations in *Analysis of Final State Implementation Plant–Rules and Regulations,* EPA Report APTD–1334 (July 1972). The

Table 8–1
Typical Emissions Factors for U.S. Incinerators in 1968 (Thousands of tons per year)

	1968 Estimate		*Lbs/Ton of Refuse*	
Pollutant	*Furnace*	*Stack*	*Furnace*	*Stack*
Mineral Particulate	90	56	15.1	9.5
Combustible Particulate	38	32	4.6	4.1
Carbon Monoxide	280	280	34.8	34.8
Subtotal (particulates)	182	142	19.7	13.6
Hydrocarbons	22	22	2.7	2.7
Sulfur Dioxide	32	32	3.9	3.9
Nitrogen Oxides	26	22	3.0	2.6
Hydrogen Chloride	8	6	1.0	0.8
Volatile Metals (lead)	0.3	0.3	0.03	0.03
Polynuclear Hydrocarbons	0.01	0.005	0.005	0.0032

Source: Niessen *et al. Systems Study . . . ,* p. I–5.

report is arranged by type of emission as well as by type of fuel (or waste) being burned. Figure 8-1 illustrates this regulatory trend.

Another concept important to an understanding of air pollution is the method that is commonly used to evaluate visible emissions from smoke stacks. Although these visible emissions may be less noxious than accompanying invisible emissions, they are more noticeable to the public. Grey or black smoke is judged by the so-called Ringlemann Chart on a scale from 0 (which indicates a clear emission) to 5 (which is completely black). White plumes are judged by opacity readings, and range from 0 to 100 density. These evaluations are becoming more and more stringent and zero-visible emissions are anticipated in the future.

In a more technical vein, "An Evaluation of Current Incinerator Emission Standards," by John R. Dervay, Han Liu, and Gregory Theoclitus, in the *Proceedings of the National Incinerator Conference* (1972), deals with the fact that there are a number of ways to specify emission control standards. These range from grains particulate per cubic foot of dry flue gas corrected to 12 percent CO_2, to grains particulate per cubic foot of dry flue gas corrected to 50 percent excess air, or pounds particulate per hundred pounds of waste

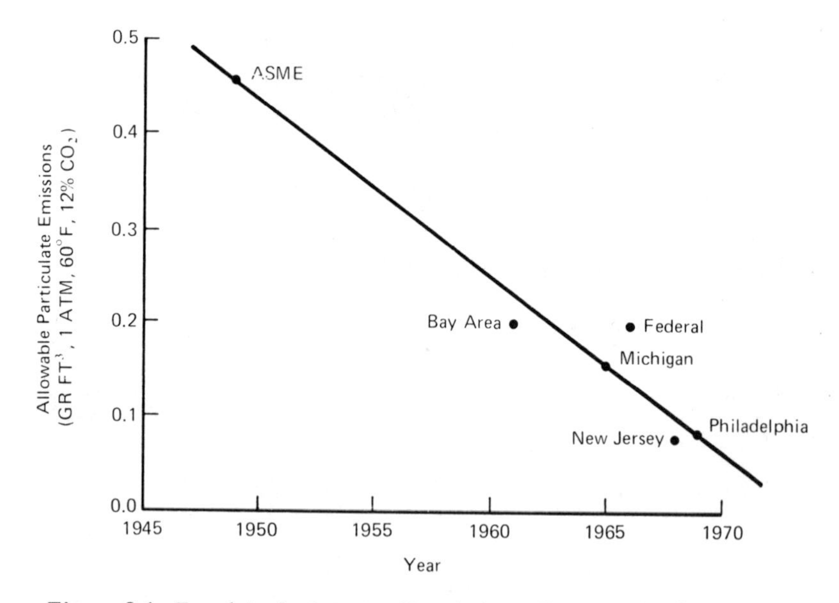

Figure 8-1. Trend in Incinerator Regulations. Source: Frank L. Cross, *Handbook on Incineration*, Westport, Conn: Technomic Publishing Company, Inc., 1972, p. 40. Reprinted by permission of Technomic Publishing Company, Inc.

charged. The paper discusses the advantages of each method and shows that there is no exact correspondence among the three. Nevertheless, a technique for converting between any one of these methods and another is shown to be both possible and useful.

Particulate Matter

When refuse is burned, fine ash particles and flakes are released. Where combustion is incomplete, carbonaceous particles are also emitted. Altogether, this particulate matter is a mixture of mineral ash, carbonaceous solids, and similar materials.

Incinerators built today and in recent years cope with these problems by high-combustion efficiency, control of combustion air, and dust collection equipment, notably electrostatic precipitators and gas scrubbers. Older incinerators are being refitted with high-efficiency (90 to 99 percent) dust collectors as part of modernization programs.

Regulations of the federal Environmental Protection Agency require that *new* incinerators of over 50-ton-per-day capacity which burn at least 50 percent municipal solid waste must limit particulate emissions to 0.08 grams per SCE corrected to 12 percent carbon dioxide (CO_2), maximum two-hour average. A final emission of 1.8 pounds of dust (particulate/fly ash) per ton of average refuse, is within these limits. A white plume of water vapor from the stack, as when caused by a gas scrubber, is not air pollution.

Dust collector technology has generally been ahead of legal requirements, but economic considerations have delayed the adoption of the most efficient equipment. Technical problems with corrosion and erosion of dust collectors and lack of a full appreciation of the care needed for the equipment have also hampered progress. Today the cost of the dust collectors and auxiliaries may represent 20 to 30 percent of the cost of the mechanical equipment and furnaces for an incinerator plant.

In general, particulate matter can be captured in two ways: by electrostatic precipitators or in water by scrubbers. In the second method, the wet dust is settled out in lagoons or water clarifiers. Part of the waste water is available for quenching grate residue; the remainder may be discharged to sewers. When incinerators are located near sewage treatment plants, sewage effluent water is available for use in the scrubbers.

Noxious Gases

Over 99.9 percent of the gases from the incinerator stack are normal constituents of the atmosphere: water vapor, carbon dioxide, oxygen, and nitrogen.

The combined total of the others normally does not exceed 0.077 percent. Nevertheless, there are a few noxious gases that largely make up that small percentage; they are discussed over the next paragraphs.

Carbon Monoxide (CO)

Carbon monoxide is a toxic, combustible gas—the result of incomplete combustion. Among other things, it is part of automobile exhaust and is produced in the smoking of tobacco and by most fires. Large quantities at low concentration are even emitted by the oceans. Although CO is absorbed by micro-organisms in the soil, man can tolerate only a small concentration because CO damages blood hemoglobin.

Fortunately, incinerator furnaces will burn CO to a negligible level by flame turbulence and ample combustion time and temperature. The ultimate desired product is carbon dioxide, CO_2. CO is apparently not removed or reduced by dry or wet dust collectors.

Nitrogen Oxides (NO and NO_2)

Some of the nitrogen in protein and other refuse components burns to nitric oxide, NO. Also, at higher temperatures in the fuel bed, some atmospheric nitrogen is oxidized to NO. However, the concentration is low compared to that from automobiles and power boilers fired by fossil fuels because incinerators operate at lower temperatures. Figure 8-2 shows what the nitric oxide equilibrium constant is as a function of temperature. Note that for the newer slagging incinerators, where the temperatures are on the order of 3000 degrees F., nitric oxide formation could be a serious problem. When NO enters the atmosphere, it is oxidized to brownish NO_2, which combines with water to produce nitric acid, H_2NO_3.

Sulfur Dioxide (SO_2)

Municipal refuse contains 0.1 to 0.2 percent of sulfur combined with other elements in almost all components of organic waste. This is less than one-tenth the sulfur content of most coals and residual fuel oils. When refuse is burned, a minor fraction of the sulfur remains with the ash, while most of the sulfur turns to the gas, sulfur dioxide. After the gases have been cooled and cleaned, some of the sulfur is found in the collected ash as well as in the emitted ash. Nevertheless, a small amount of sulfur dioxide is present in the stack gases.

Sulfur dioxide is oxidized by sunlight in the atmosphere to SO_3 and combines readily with water, returning to the soil and to plants.

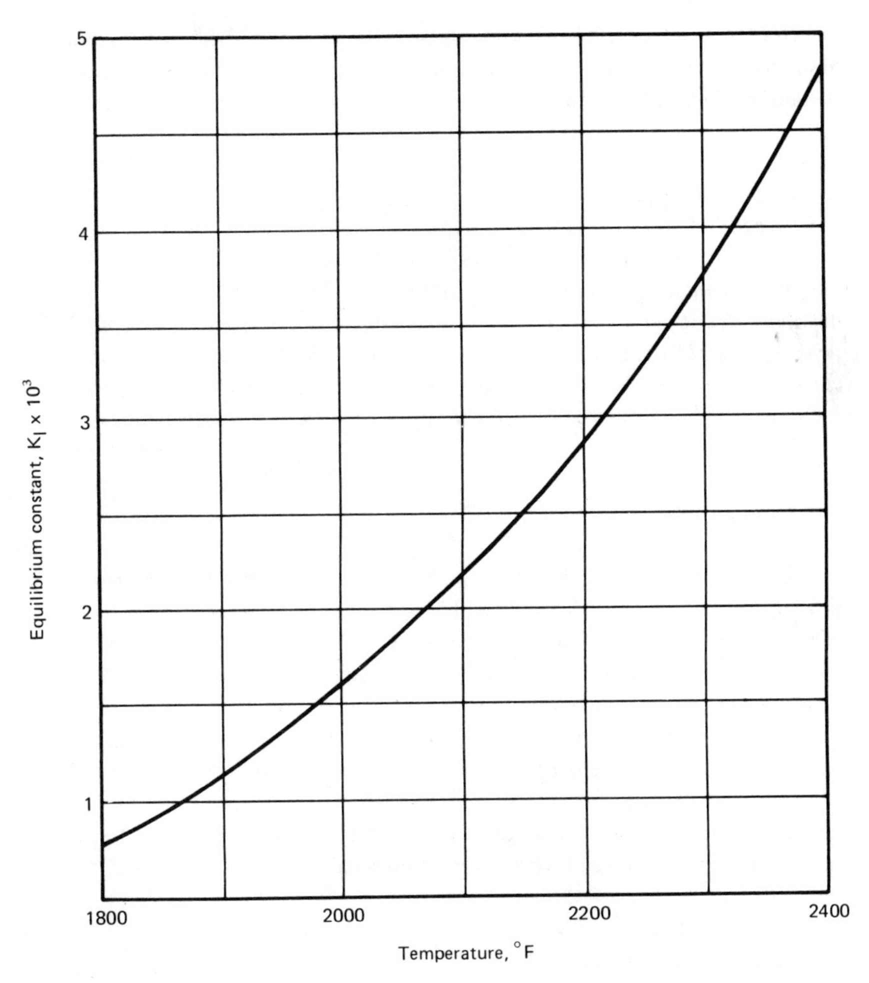

Figure 8-2. Equilibrium Constant for Nitric Oxide Formation. Source: *Principles and Practices of Incineration*, edited by Richard C. Corey, p. 30. Copyright© 1969 by John Wiley & Sons, Inc. Reprinted by permission of the publisher.

Hydrogen Chloride (HC1)

Almost all refuse organic matter contains some combined chlorine–some as common salt (NaC1) and some in chlorinated plastics, such as polyvinyl chloride. Fire conditions cause the formation of hydrogen chloride, a gas which forms hydrochloric acid with water. Part of the HC1 recombines with lime and other mineral dusts in the cooler parts of the incinerator system and is trapped at about 90 percent efficiency in wet scrubbers.

HC1 is apparently an unavoidable product of incineration. Although it has a negligible effect on incinerator grates and refractories, precautions must be taken to prevent or limit its corrosive effect on boiler tubes, scrubbers, piping and pumps.[1]

Organic Acids

When refuse organic matter is heated in the fire in zones where air is deficient, acetic acid (CH_3COOH) and organic acids are among the combustible gases evolved. Most of the organic acid is burned above the fuel bed, but minor amounts in the parts-per-million range escape. Furnace temperatures over 1500 degrees F. and turbulence in the furnace promote burnout of these gases.

Odors

Combustion destroys most odors. Stack gas from well-run municipal incinerators has a negligible odor, which is quickly brought below the threshold of perception by dilution in the atmosphere. Odors from refuse in the storage pit and the unloading operation can be minimized by housekeeping and clean-up around the plant. It is only important that moist refuse not be stored longer than necessary because odors can be produced by bacterial action. Withdrawal of air from the enclosure over the storage pit also helps prevent odors from escaping.

Odors from incinerator residue are minimized by good burnout, followed by complete quenching before hauling to disposal.

Control of Particulates

For the most part, there are three factors of incineration that result in high emission rates:

1. The mechanical entrainment of particles from the burning refuse bed.
2. The cracking of pyrolysis gases.
3. The volatilization of metallic salts or oxides.

The first of these results from refuse with a high percentage of fine ash, by high underfire air velocities, or by other factors that induce a high gas velocity in the furnace and through the bed. The second comes from refuse with a high volatile content producing pyrolysis gases with a high carbon content and by conditions above the fuel bed that prevent the burnout of the coked particles formed by the cracking of the volatiles. The third results from high concentration of metals that form low-melting point oxides and by high temperatures within the bed.

As for particulate emission, Figure 8-3 shows the actual emission rates per ton of refuse for many operating incinerators. Naturally, the wide variation results from the many furnace designs and sizes that were studied. Still, the amount of underfire air is another important factor. Some, but not all, of the mineral particulate may be eliminated by reduction of the rate of underfire air. Certain minimum levels of underfire air are necessary, however, to cause

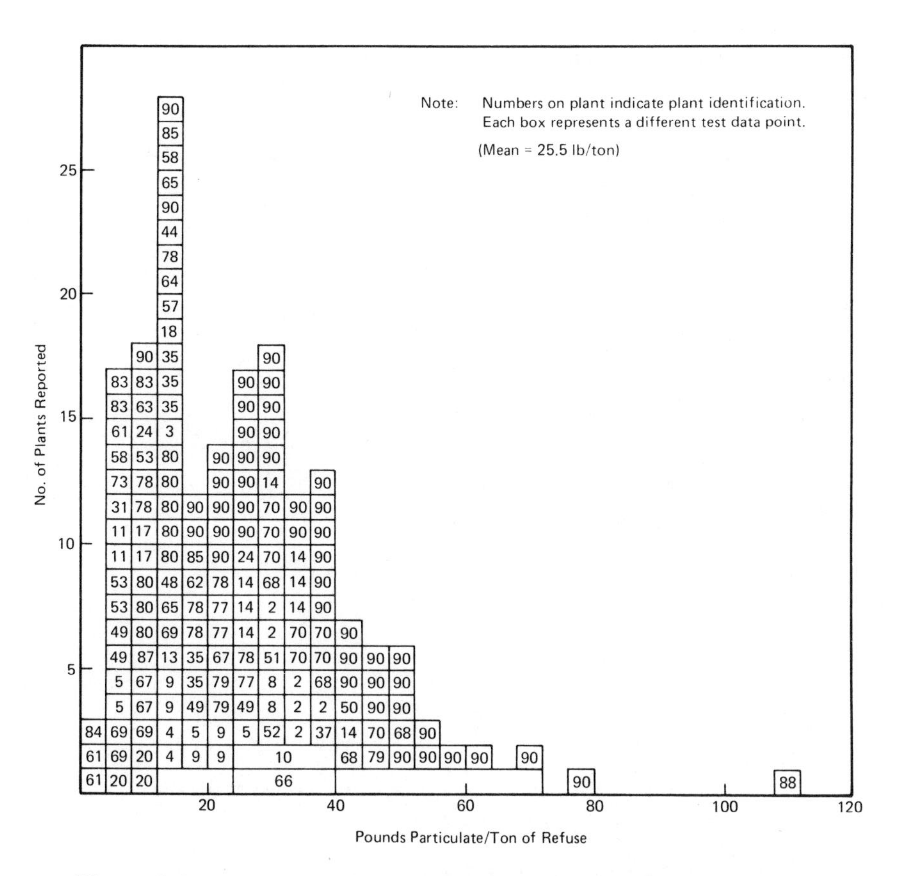

Figure 8-3. Histogram of Particulate Furnace Emission Factors for Municipal Incinerators (Capacity: < 50 ton/day). Source: W.R. Niessen and B. Sarofim, "Incinerator Air Pollution: Facts and Speculation," *Proceedings of the National Incinerator Conference* (1970), ASME, p. 170.

efficient burnout on the grate. Too great a reduction in underfire air can therefore reduce the tonnage through-put and increase the unburned portion of the residue. Figure 8–4 shows the effect of underfire air on emission rates at five furnaces.

The following properties of particulate emissions are considered important from the standpoint of their effect on the control and collection of the pollutants: quantities, particle size distribution, specific gravity, electrical characteristics, and chemical composition.[2] Generally, it is easier to collect large particles that have a high specific gravity. Fine, light materials require more sophisticated techniques.

Niessen and Sarofim, writing in "Incinerator Air Pollution: Facts and Speculation," have made the following summary of their conclusions concerning particulate air pollution from municipal incinerators:

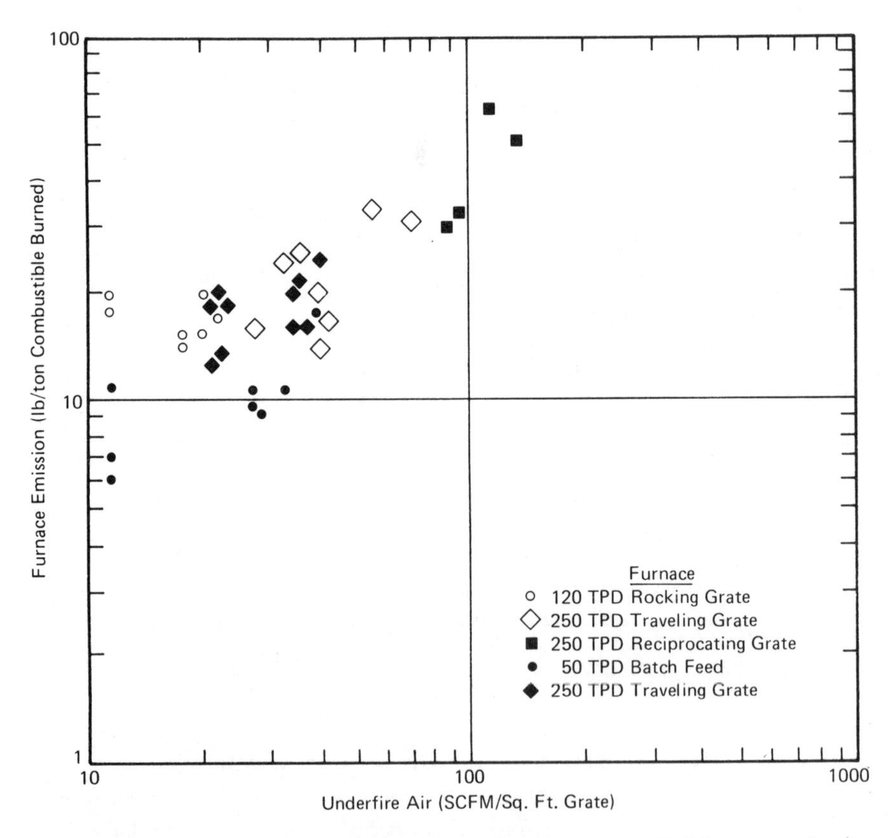

Figure 8-4. Entrained Particulate Emissions. Source: DeMarco *et al.*, *Incinerator Guidelines–1969*, p. 51.

> The average total (mineral and combustible combined) particulate emission factor is 24 pounds per ton of refuse, although wide variations exist.
>
> The mineral particulate loading is nearly in direct porportion to the quantity of ash (excluding metal and glass) in the refuse for equivalent designs and operations. The combustible particulate loading is strongly related to the volatile content of the refuse (perhaps more closely to the volatile carbon content).
>
> Combustible particulate persistence seems related to furnace overfire mixing processes rather than combustion-rate control.
>
> Underfire air rate ($ft^3/min/ft^2$) seems strongly related to mineral-particulate emission rate. Particulate emissions seem related to furnace and stoker type in the order: cylindrical-batch < traveling-continuous < rocking-continuous < rectangular-batch < reciprocating-continuous.[3]

The earliest attempts at particulate pollution control consisted merely of settling chambers, which, as the name implies, were chambers through which the gas passed after leaving the furnace, and where it remained just long enough for gross particulate matter to settle out. Reported efficiency of settling chambers ranges from 34 to 60 percent, not sufficient for present-day requirements.[4]

Wetting baffle systems have been widely used in municipal incinerators, being installed in over half of the new plants constructed since 1957. These consist of one or more baffle screens sprayed with water. Efficiency varies widely, but ranges between 10 and 50 percent.

Cyclone collectors have been installed on about 20 percent of the plants built in the United States since 1957. Little information is available concerning their performance, but maximum efficiency is stated to be approximately 70 to 80 percent.

The use of fabric filters is now in the experimental stage in municipal incineration. Efficiencies are very high, approaching 100 percent particle removal. The dust cake that forms on the fabric does most of the filtering work, and its high efficiency results from the fact that the pore openings in the cake are no bigger than the smallest particle in the gas stream. There is no reliable information as yet concerning bag life or operating and maintenance costs, but it is felt by some that cost considerations will prevent the use of fabric filters for incineration. In addition, unless conditions are closely controlled, blinding can be a problem and the acid in the flue gas can attack the fabric, which can make for unreasonably high maintenance costs.

The two types of air pollution control equipment now receiving the most attention in the United States are the wet scrubber and the electrostatic precipitator. Both systems are available at various efficiencies, with costs increasing with efficiency.

To meet today's air pollution regulations, over 90 percent of the particulate matter that enters the collectors must be captured. Efficiencies up to 99 percent are available. Table 8-2 shows the efficiency of many types of APC devices. Electrostatic precipitators increase in size and cost to meet the higher efficiencies. Wet scrubbers meet higher efficiencies by increasing the mixing of water and gas at the expense of higher fan horsepower.

A paper by J.H. Fernandes details the operation of both types of equipment.[5] In the case of scrubbers, several facts are known. First, the dust particle must hit the water droplet. The efficiency of this impact is a direct function of velocity and an inverse function of the diameter of the droplet. The unit's collection efficiency is therefore a direct function of the power supplied to the unit. Two types of scrubbers are shown in Figure 8-5. In the high-velocity cyclonic spray scrubber, the dust-laden gases are introduced at high velocity and are impinged with a spray from a central manifold. The circulating gas rises through this curtain of water; as the droplets pick up dust particles, they are also driven to the wall and recirculated. At the neck of the unit, the gas is treated by anti-spin vanes and a mist eliminator.

In the packed scrubber, shown in Figure 8-5, the gas passes through layers of materials that are constantly being sprayed. As the gases make their way to the top of the system for exhaust, noxious gases and particles are absorbed.

In a venturi scrubber, shown in Figure 8-6, a high-velocity gas stream is formed at the throat of the venturi where the water is introduced. The gas stream itself breaks the water down into fine droplets.

Electrostatic precipitators are also straightforward (Figure 8-7). In these devices, the dust particles in the gas stream are electrostatically charged by a high-voltage discharge and after charging pass into a high voltage electrostatic field where the particulate matter is attracted to a negatively charged collection surface. From that surface, the dust particles can either be shaken or hosed down at intervals.

Since optimum operating temperatures for the electrostatic precipitators are between 470 and 520 degrees F., the furnace gases must first be cooled to this range by heat absorption in a boiler or spray chamber. By contrast, scrubbers can accept gases at any temperature in a quench section ahead of the scrubber. The gases cool only slightly in the precipitator, but leave the scrubber at about 160 to 170 degrees F., saturated with water vapor. Wet scrubbers have the added advantage of being able to remove certain water-soluble gaseous pollutants. They pose the problem, however, of disposal or cleaning of the contaminated water, and of the aesthetic and psychological impact of the vapor plume emitted from the stack.

In regard to the just-mentioned temperature problems, in "Flue Gas Cooling," by V. Westergaard and J.A. Fife, presented at the ASME 1964 National Incinerator Conference, the authors discuss the problem of dealing with the fact that typical air pollution control equipment will not operate on the gases

Table 8–2
Average Control Efficiency of APC Systems—APC System Removal Efficiency (Weight Percent)

APC Type	*Mineral Particulate*	*Combustible Particulate*[a]	*Carbon Monoxide*	*Nitrogen Oxides*	*Hydro-carbons*	*Sulfer Oxides*	*Hydrogen Chloride*	*Polynuclear Hydro-carbons*[b]	*Volatile Metals*[c]
None (Flue Setting Only)	20	2	0	0	0	0	0	10	2
Dry Expansion Chamber	20	2	0	0	0	0	0	10	0
Wet Bottom Expansion Chamber	33	4	0	7	0	0	10	22	4
Spray Chamber	40	5	0	25	0	0.1	40	40	5
Wetted Wall Chamber	35	7	0	25	0	0.1	40	40	7
Wetted, Close-Spaced Baffles	50	10	0	30	0	0.5	50	85	10
Mechanical Cyclone (dry)	70	30	0	0	0	0	0	35	0
Medium-Energy Wet Scrubber	90	80	0	65	0	1.5	95	95	80
Electrostatic Precipitator	99	90	0	0	0	0	0	60	90
Fabric Filter	99.9	99	0	0	0	0	0	67	99

[a] Assumed primarily < 5 microns.

[b] Assumed two-thirds condensed on particulate, one-third as vapor.

[c] Assumed primarily a fume < 5 microns.

Source: Niessen *et al.*, *Systems Study* . . . , p. III–34.

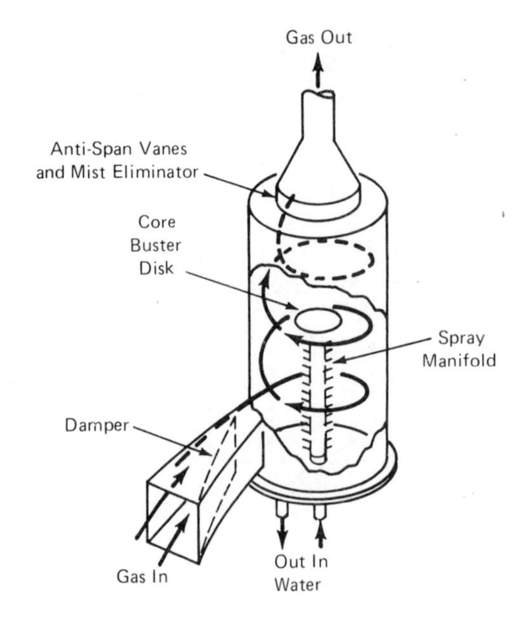

Cyclonic Spray Scrubber.

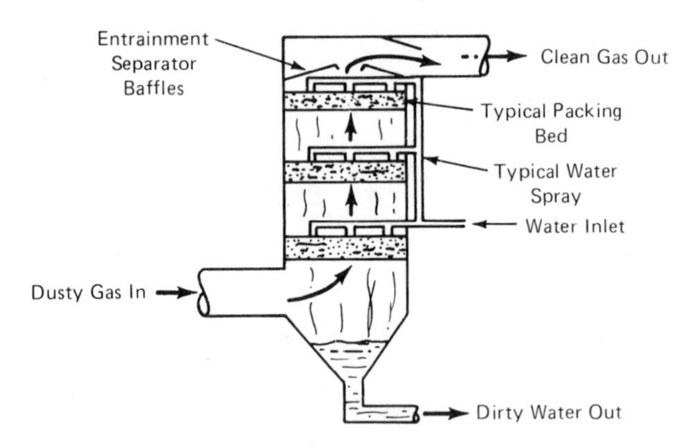

Packed Scrubber.

Figure 8-5. APC Devices–Cyclonic Spray Scrubbers and Packed Scrubbers. Source: J.H. Fernandes, "Stationary Source Air Pollution Control Techniques and Practices in the United States," presented at the Industrial Air Treatment and Pollution Control Equipment Symposium, Frankfurt, Germany, November 10-12, 1970, p. 7.

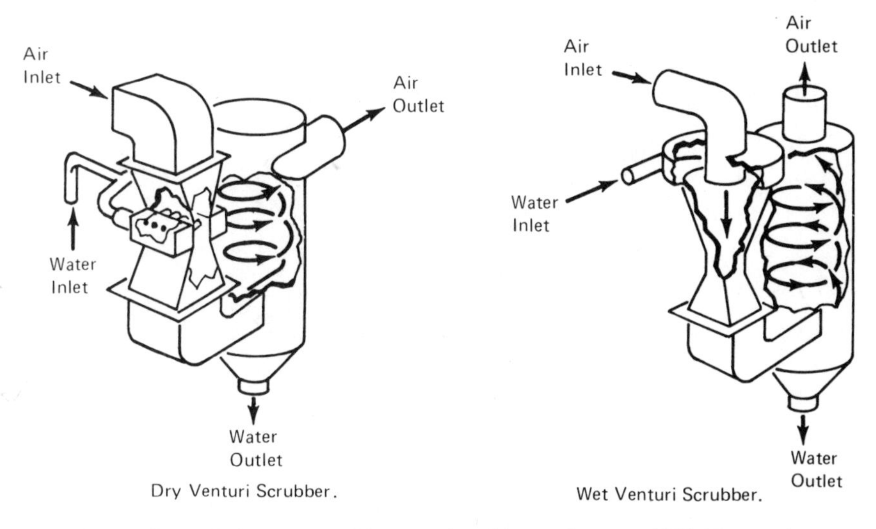

Figure 8-6. APC Devices–Venturi Scrubbers. Source: J.H. Fernandes, "Stationary Source Air Pollution Control Techniques and Practices in the United States," presented at the Industrial Air Treatment and Pollution Control Equipment Symposium, Frankfurt, Germany, November 10-12, p. 8.

when temperatures are in excess of 600 to 700 degrees F. (Mechanical collectors and precipitators can and do handle gases above 600 to 700 degrees F., but the volumes are excessive and the materials of construction are inordinately expensive.) Essentially, the designer has three means for reducing the temperatures of the hot combustion gases: (1) he may dilute the gases with ambient air; (2) he may quench the gases by spraying water into them, which evaporates and reduces the combined gas temperature; or (3) he may design a heat-transfer surface such as a water-wall incinerator, where the primary heat of combustion is transferred outside of the gases to the colder walls.

Figure 8–8 shows what the gas volume is as a function of method of cooling the gas from 1500 to 700 degrees F. As can be seen, the maximum volume of hot gases entering the air pollution control equipment occurs when the gases are quenched with cold air. The minimum volume of hot gases entering the air pollution control equipment occurs when the heat is transferred outside through water-wall construction.

Table 8–3 shows, in tabulated form, the volumes of air and water necessary to quench the combustion products. It must be kept in mind, though, that water quench, while it reduces the cost of air pollution control equipment over that of using tempering air, will require the expense of installing plumbing and water treatment equipment. In addition, there will still be an unsightly steam

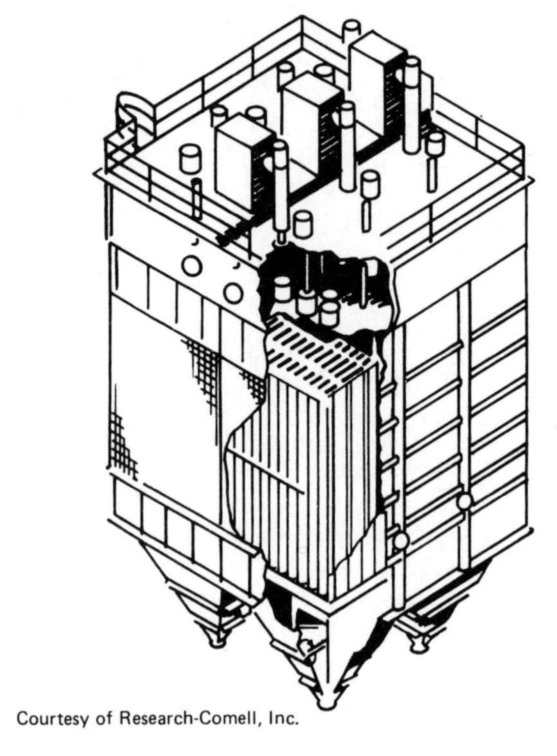

Cut away of typical precipitator.

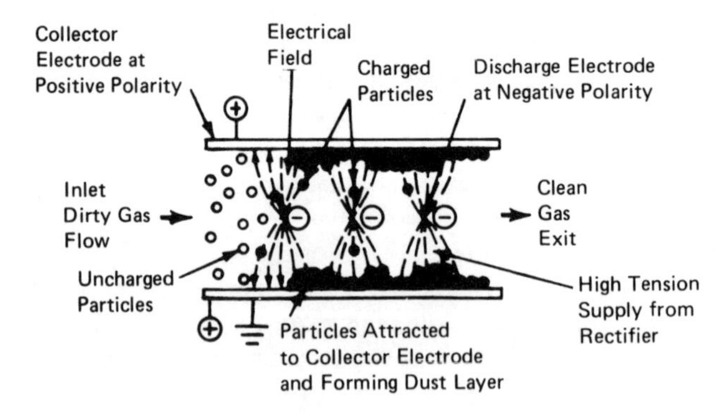

Figure 8-7. APC Devices–Electrostatic Precipitators. Source: J.H. Fernandes. "Stationary Source Air Pollution Control Techniques and Practices in the United States," presented at the Industrial Air Treatment and Pollution Control Equipment Symposium, Frankfurt, Germany, November 10-12, p. 9.

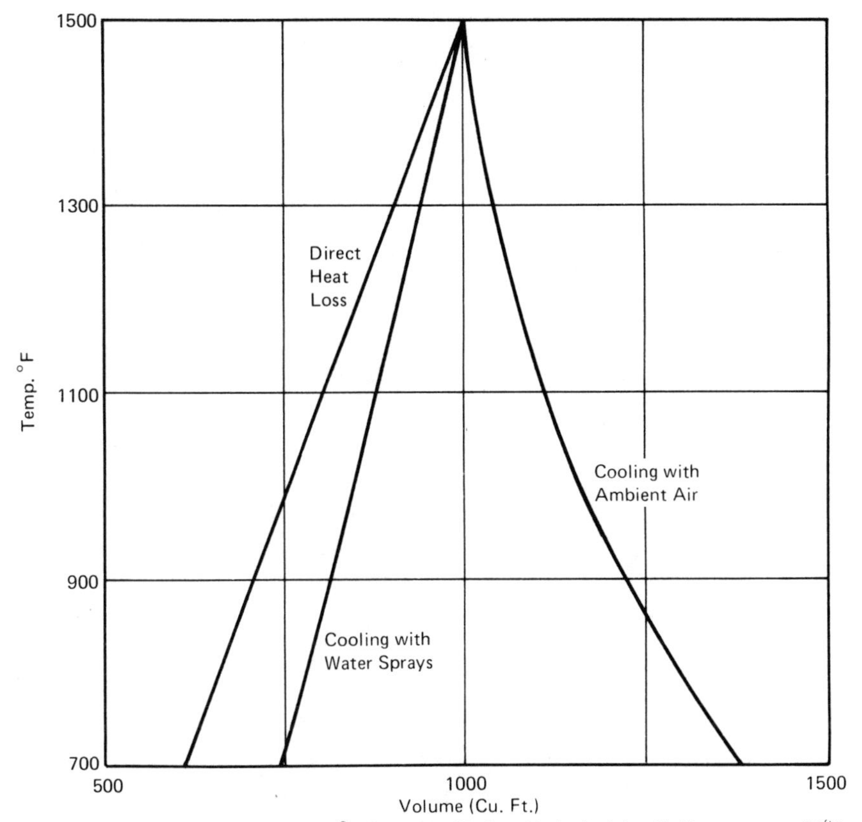

Volume of 1000 Cu. Ft. Gases at 1500°F Cooled by Various Methods. Inlet Air Temperature -80°F; Inlet Water Temperature -70°F. Based on 100% Efficiency.

Figure 8-8. Volume Changes of 100 Cubic Feet of Gases from 1500°F., as Cooled by Various Methods. Source: V. Westergaard and J.A. Fife, "Flue Gas Cooling," *Proceedings of the National Incinerator Conference* (1964), ASME, p. 171.

plume. The cost of water-wall construction is also more expensive than ceramic-wall construction in the furnaces, but maintenance costs are significantly less.

"Characterization of Several Incinerator Process Waters," by D.A. Wilson and R.E. Brown, in the *Proceedings of the National Incinerator Conference* (1970), discusses the contaminants in the incinerator waters from both fly ash and spray water chambers and from quench tanks. Another paper dealing with the subject is "Characterization and Treatment of Incinerator Process Waters," by R.J. Schoenberger, P.W. Purdom, S.S. Levey and H.I. Hollander, also in the 1970 *Proceedings.*

Although air pollution control equipment has been developed to clean stack emissions, the costs are high. Figures 8-9 and 8-10 present the investment

Table 8–3
Volume Changes of 1000 Cubic Feet of Gases from 1500°F., as Cooled by Various Methods

Cooling by Direct Heat Loss[a]		*Cooling by Ambient Air Based on 100 Per Cent Efficiency*				*Cooling by Water Sprays Based on 100 Per Cent Efficiency*			
To Temp. F.	*Vol. Cu. Ft.*	*Lb. Entering Gas*	*Required Dilution Air Lb. at 80°*	*Total Lb. Gases Leaving*	*Total Volume Cu. Ft.*	*Lb. 70° Water Required*	*Total Cu. Ft. Dry*[b] *Gases*	*Total Cu. Ft. Vapor*	*Total Cu. Ft. Gases & Vapor*
1500	1000	20.20	0.00	20.20	1000	0.00	1000	0	1000
1300	898	20.20	3.57	23.77	1054	0.59	898	42	940
1100	796	20.20	8.57	28.77	1130	1.25	796	79	875
900	694	20.20	16.01	36.21	1233	1.97	694	109	803
700	592	20.20	28.20	48.40	1415	2.88	592	135	727

[a]Cooling by supplying heat to a waste heat boiler or similar equipment.

[b]By dry gases are meant the combustion-products containing the moisture of the refuse but not including moisture added by the cooling process.

Source: V. Westergaard and J.A. Fife, "Flue Gas Cooling," *Proceedings of the National Incinerator Conference* (1964), ASME, p. 171.

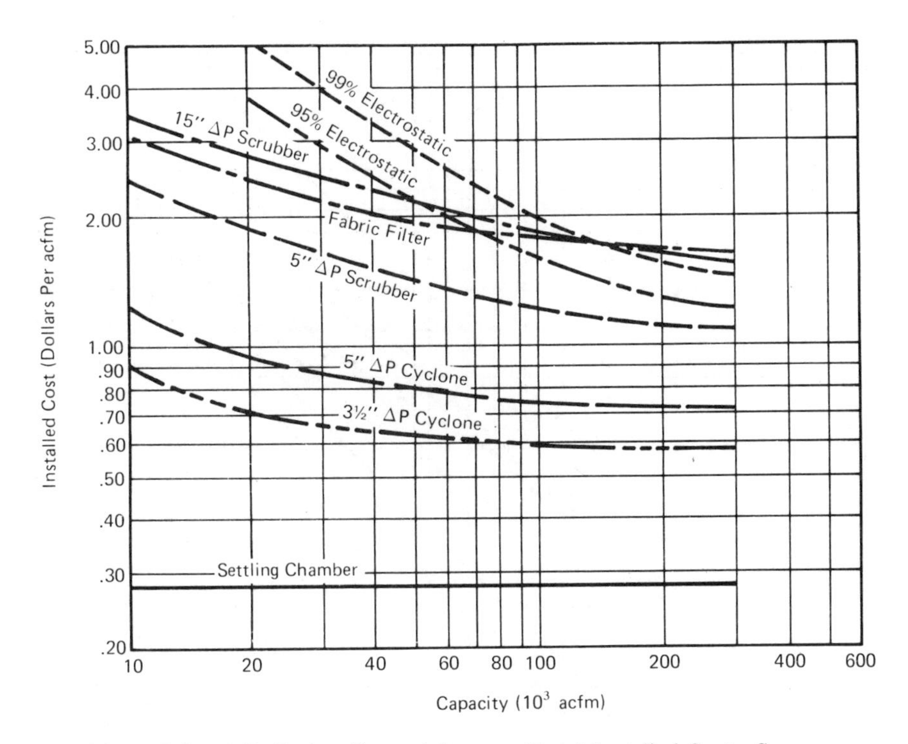

Figure 8-9. Air Pollution Control Systems Total Installed Costs. Source: Niessen *et al., Systems Study . . .*, p. III-35.

and operating costs for the various air pollution control devices. Figure 8-11 presents the trade-off between efficiency and operating cost.

Costs

Electrostatic precipitators are slightly more expensive to install and operate than wet scrubbers. Table 8-4 shows the effects of the two systems on the capital costs of an 800-ton-per-day incinerator in the District of Columbia. Table 8-5 lists comparative operating costs for the alternative pollution control systems for the same plant.[6]

Emission Testing

Proof of the adequacy of the air pollution control equipment, together with drawings, must be submitted to control agencies for approval before the

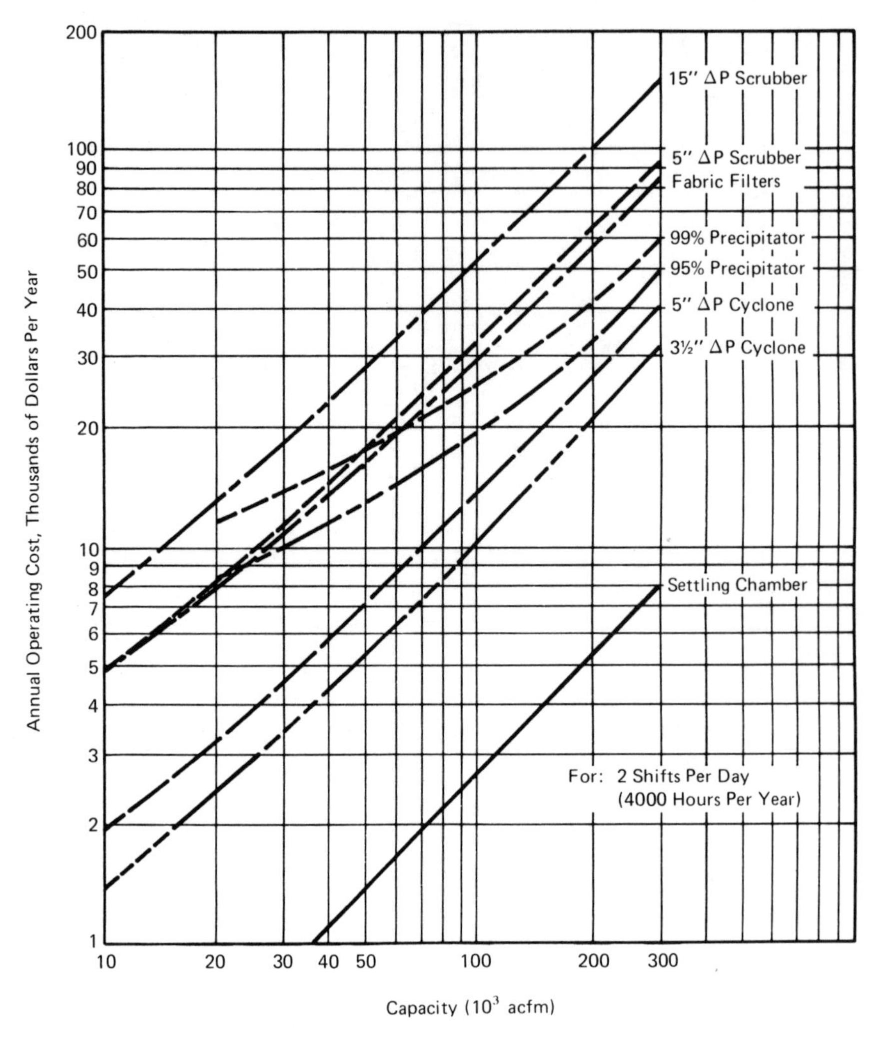

Figure 8-10. Air Pollution Control Systems Annual Operating Costs. Source: Niessen *et al., Systems Study* . . ., p. III-36.

start of construction of new incinerators or major remodeling of existing plants. Soon after the plant is in operation, a test must be conducted to measure the opacity of the smoke from the stack and concentration of the particulate emission in the stack gases during rated operating capacity. A number of firms perform such tests on a fee basis.

Scaffolding is erected to ports part way up the stack. Probes are inserted

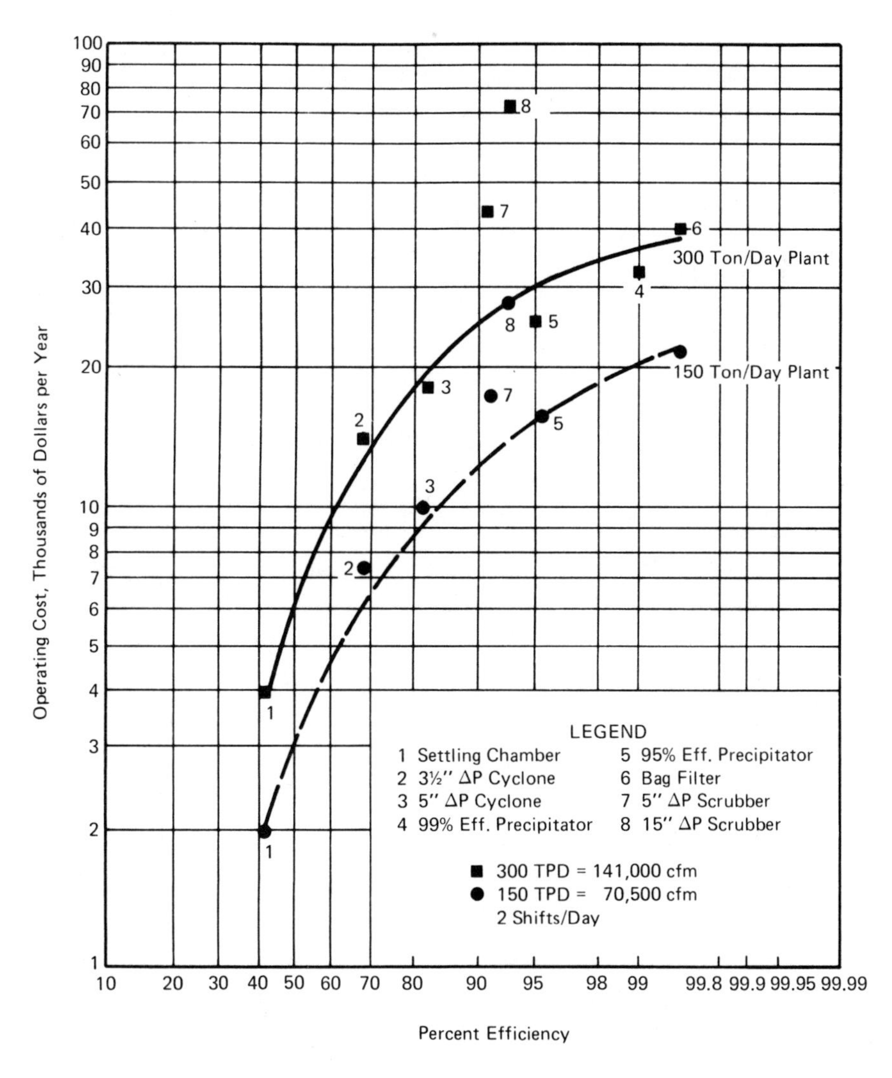

Figure 8-11. Total Annual Operating Cost Versus Particulate Removal Efficiency. Source: Niessen *et al., Systems Study . . .*, p. III-37.

into the stack to measure the gas temperature and to withdraw a small stream of the gas at the same velocity as in the stack (isokinetic sampling). The gas is filtered to trap the solids, after which the solids are carefully weighed.

The data are reported, together with the weight of particulate matter per unit volume or weight of dry flue gas, corrected to 12 percent carbon dioxide

Table 8-4
Comparative Capital Cost Estimates, Selected Items for Air Pollution Control Study, Four-Unit Incinerator Plant

	Type of Air Pollution Control Equipment	
	Electrostatic & Mechanical	*Wet Scrubber*
General building contract		
Incinerator building & foundations	$ 606,700	$ 447,000
River pump house	–	7,200
Clarifier basins	–	45,000
	$ 606,700	$ 499,200
Mechanical contract		
Refractory furnaces and flues	$ 204,000	$ 248,400
Spray cooling chamber	221,900	–
Steelwork	324,000	228,000
Insulation	133,700	26,500
Instrumentation	60,000	42,000
Installation of purchase equipment	132,000	60,000
Piping	–	58,500
	$1,075,600	$ 663,400
Purchased equipment		
Fans and drives	$ 135,900	$ 125,900
Pumps and drives	–	19,000
Air pollution control equipment	396,000	337,300
Clarifier equipment	–	64,800
	$ 531,900	$ 547,000
Electrical contract: Power & lighting	$ 195,000	$ 129,000
Subtotal: Physical cost	$2,409,200	$1,838,600
Engineering and field supervision	169,000	130,000
Contingency	241,000	185,000
Escalation to December 1968	120,500	92,900
Total incremental physical	$2,939,700	$2,247,400

Source: Day and Zimmermann, *Special Studies for Incinerators,* p. 18.

content at the standard temperature and atmospheric pressure. The requirements vary with the states. Reports indicate the legal limits and whether or not the plant complied with the requirements.

Corrosion

Another problem with incineration operations is the fact that several system components are subject to chemical attack—corrosion. While corrosion of incinerator grates is not a serious problem—and refractories are also more

Table 8-5
Overall Operating Costs for Electrostatic Precipitators and Wet Scrubbers (Annual Costs for 800 Ton/Day Plant)[a]

	Electrostatic	*Scrubber*
Maintenance	$139,000	$127,700
Electric Power	160,600	130,800
Purchased Water	27,900	1,000
Subtotal	$327,500	$259,500
Fixed Charges (20 years; 4½%)	185,000	141,500
Total	$512,500	$401,000
Total, $/Ton at Capacity	1.77	1.38
Total, $/Ton at 70% Capacity	2.50	2.00

[a]Costs comparable but incomplete.
Source: Day and Zimmermann, *Special Studies for Incinerators,* p. 17.

or less resistant—high-temperature corrosion must be guarded against on the furnace, superheater, and boiler tubes. Low-temperature corrosion can be a problem on economizers, air heaters, wet gas scrubbers, and electrostatic precipitators. Corrosion can also occur on the water side of a wet gas scrubber in such areas as waste water piping and pumps.

P.D. Miller *et al.*, of Battelle Memorial Institute, investigated metal-wastage rates in water-wall incinerators and found that the attack of tube metal increased with temperature in the high-temperature areas of the unit:

> The contributors to the attack are corrosive gases (SO_2, SO_3, HCl, and Cl_2) and low-melting chloride and sulfur-containing salts which exert a fluxing action on the protective films on the metal surface. These low-melting salts primarily contain compounds such as zinc and lead chlorides along with potassium bisulfate and potassium pyrosulfate."[7]

In addition to these gases, Miller states that the amount of corrosion is also influenced by such things as:

1. Whether the metal is exposed to reducing or oxidizing atmosphere.
2. The amount of volatilized inorganics that deposit on the tubes.
3. How often the outside scale is removed from the tubes.[8]

From this it is very easy to gather that high-temperature corrosion is a very complicated phenomenon.

In general, it has been found advisable to protect boiler tubes in critical areas low in the furnace with refractory coatings held in place by tube studs. Combustion must be controlled to avoid impingement of combustible gases on the boiler tubes. A study by Karl Thoemen summarizes tube-corrosion experience over an operational period of six and one-half years at the Dusseldorf incinerator, one of the first water-wall steam-generating incinerators.[9]

It should be pointed out that adding refractory to the tubes to prevent corrosion is a corrective action that could hopefully be avoided.

The Paris Ivry Incinerator Plant operated the boiler at 1380 psig at the superheater outlet, which meant the boiler's operating pressure was probably around 1500 psig. which corresponds to roughly 600 degrees F. steam pressure in the water-wall tubes and approximately 650 degrees F. outside metal temperature. This, coupled with a possible alternating oxidizing, reducing atmosphere, caused severe furnace tube corrosion. This problem was remedied by covering the tubes with refractory that, in turn, destroyed the effectiveness of the furnace to cool the products of combustion down to a point where they did not plug the boiler passes. To remedy this new problem, they had to recirculate cooler gases from the colder end of the unit.[10]

Corrosion of iron, carbon steel, and many grades of stainless steel occurs when incinerator flue gases contact cool surfaces and cause condensation of moisture with acid gases. To minimize this attack on economizers, the feedwater temperature must be held above the critical temperature, and on air heaters, the air and gas temperatures must be controlled to maintain the metal temperature above the danger point. To minimize this kind of damage on electrostatic precipitators, efforts are made to hold the entering gas temperatures above the low-temperature problem area; they are usually heat-insulated.

Gas scrubbers of metal construction have also suffered chloride corrosion. To prevent this, the use of acid brick, rubber linings, and polyester plastic with fiberglass reinforcement is advisable, as reported by Eugene Backus in *Civil Engineering.* Similarly, plastic water piping, gas ducts, and stacks are coming into use on scrubber installations.[11]

General Economic Considerations

A major disadvantage to municipal refuse incineration has long been its cost, which has grown dramatically over the past 10 years and is expected to continue to increase due to inflation, the complex nature of the new designs, and the cost of skilled labor. However, these high costs per ton processed may often be justified in the larger plants by the sale of recovered materials to be recycled or, in the case of a steam plant, through the sale of steam.

Capital Investment

The best capsule discussion of incinerator costs is to be found in *The Treatment and Management of Municipal Solid Waste,* edited by David Gordon Wilson (1972). Much of what follows is taken from that report.

To compare the economic aspects of the various incineration concepts now used, a common design basis was selected. The data that follow assume that no waste heat or metal recovery credits are taken; that refuse is fed directly from a pit to the furnace by means of a bridge crane and bucket; and that the incineration facility is on land suitable for construction without excessive site preparation or road construction.

Existing *batch-incineration systems* show a wide variation in cost. The data shown in Figure 8-11 were accumulated for the furnace and building portions of existing batch systems (both of the rectangular and cylindrical type) adjusted to reflect labor and materials costs in Boston, Massachusetts, for mid-1969. The appropriate incremental costs for electrostatic precipitators were then included to make up the total investment. Figure 9-1 shows the resulting plant costs in dollars per ton of rated capacity per 24-hour day (TPD) as a function of the capacity of furnace units comprising a plant with two or more furnaces. The range of typical furnace, building, and electrostatic precipitator costs is shown. The furnace costs include: the crane, the primary and secondary combustion chambers, forced-draft fans and ducting, residue-removal system, gas-cooling system, breeching and stack, piping, instrumentation and controls. The building costs include excavation, land improvements, foundations, the refuse pits, buildings and structural supports, and a moderate investment in access road. The electrostatic precipitator costs include the induced-draft fan

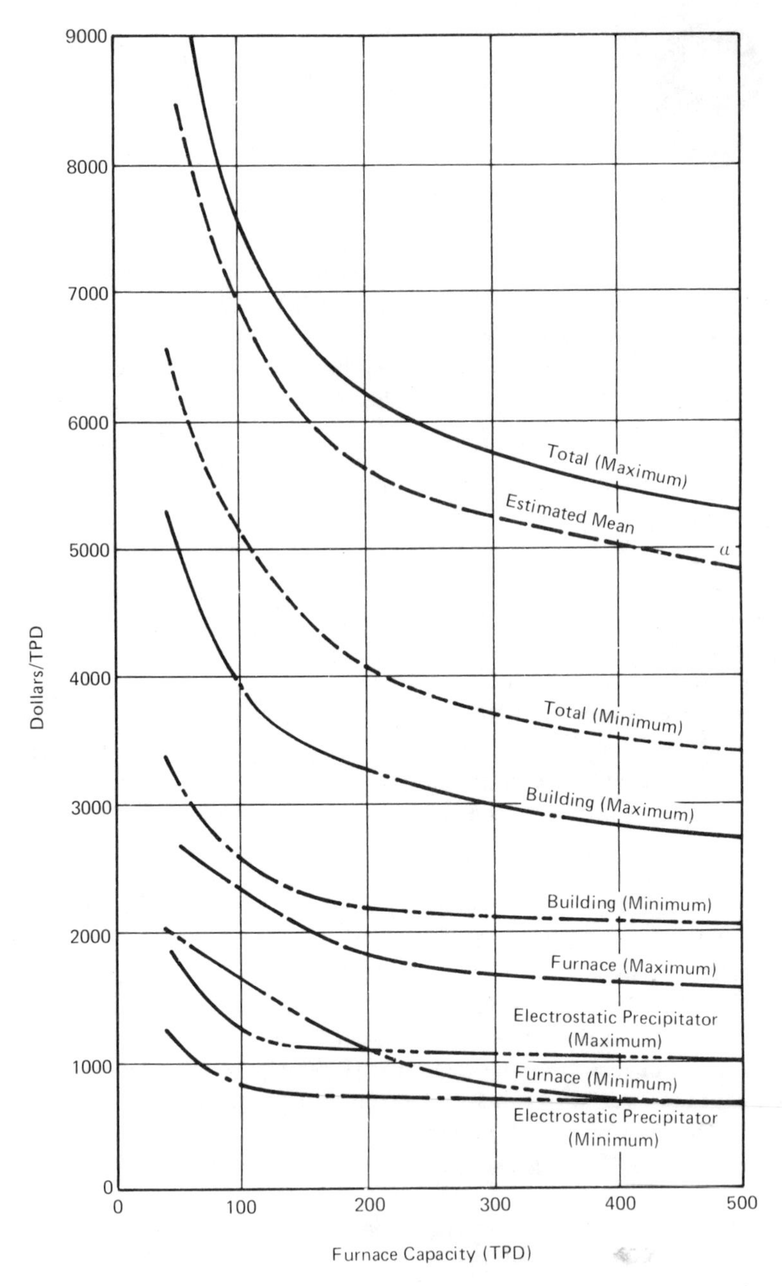

Figure 9-1. Average Investment Costs—Batch-Feed Systems. Source: Niessen *et al., Systems Study . . .*, p. VII-99.

plus all auxiliaries, motors, and ducting directly associated with the precipitator system.

The figure illustrates well the wide range of batch-fed incinerator costs. For example, an incinerator comprised of two 150-TPD units might generally be expected to have an initial capital cost of about \$6,000 per daily ton. However, because of variations in the parameters discussed above, plants of similar capacity could cost from \$4,400 to \$6,700 per daily ton.

The *continuous-feed incinerator* with water-wall construction and grate burning is similar in many respects to existing plants with refractory-wall construction. Refuse-feeding, grate, siftings-removal, and ash-handling sub-systems are or can be identical. The primary differences lie in the cost and the operating characteristics of the units.

Figures 9-2 and 9-3 show, first, the estimated investment costs for refractory construction incinerator plants, and then, a similar relationship for grate-fired, water-wall units with convection boilers and electrostatic precipitators. The values include a 10 percent override for site preparation, costs for a steel stack and a building with housed pit, but no housing over the tipping floor. Since steam demand may not exist during periods of time when the incinerator is operating, the cost for an air condenser has also been included.

Operating Costs

Reported operating costs vary widely. The false economy of deferred maintenance, or the penalties of under-designed equipment, for example, cause wide swings in published values for maintenance expense for similar plants. In some plants, efficient operation is obtained with only three or four operators whereas, in others, unreasonably large crews may be found. To compare the economic aspects of various incineration concepts, a common operating basis must be selected. The data that follow assume that no waste heat or recovery credits are taken, and also assume standard unit costs for electricity, water, labor, and overhead. A reasonable crew size was taken; one comparable with many of the better-run plants in the country.

Figure 9-4 shows the total operating costs (including capital charges) for cylindrical- or rectangular-type batch-feed incineration plants. This figure clearly shows the advantages in multi-shift operation. For example, at a 300-ton-per-day disposal rate, a plant running three shifts experiences a total operating cost reduction of almost \$2 per ton in comparison with one-shift operation and simultaneously acts to reduce furnace maintenance by holding more constant temperatures, avoiding refractory spalling, condensation and so forth. Table 9-1 shows operating economics calculated for a specific 400-ton-per-day plant to illustrate the breakdown of costs between the various cost centers.

Figure 9-5 shows costs for continuous-feed systems. It can be seen that in

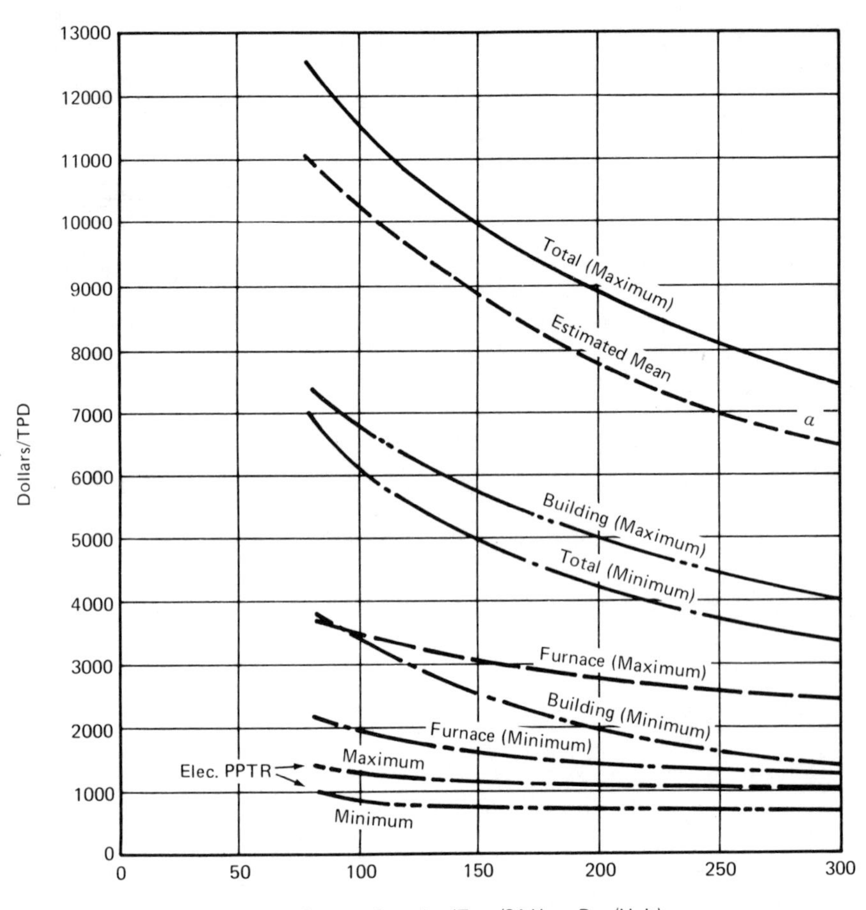

Figure 9-2. Average Investment Costs–Continuous-Feed Systems. Source: Niessen *et al.*, *Systems Study* . . ., p. III-41.

larger plants, the operating cost for continuous-feed systems is only slightly higher than that for batch-feed systems. For a minor additional cost, however, substantially improved residue quality and diminished combustible air-pollutant emissions are obtained. Yet, it can be easily seen that the unit cost for continuous systems increases rapidly as plant size diminishes.

Figure 9–6 shows the operating cost for water-wall incineration systems. Notice that only the three-shift-per-day operation is shown, as continuous operation is highly desirable with boiler systems. It can be seen, in comparison with the continuous-operation refractory systems, that water-wall operation, due to

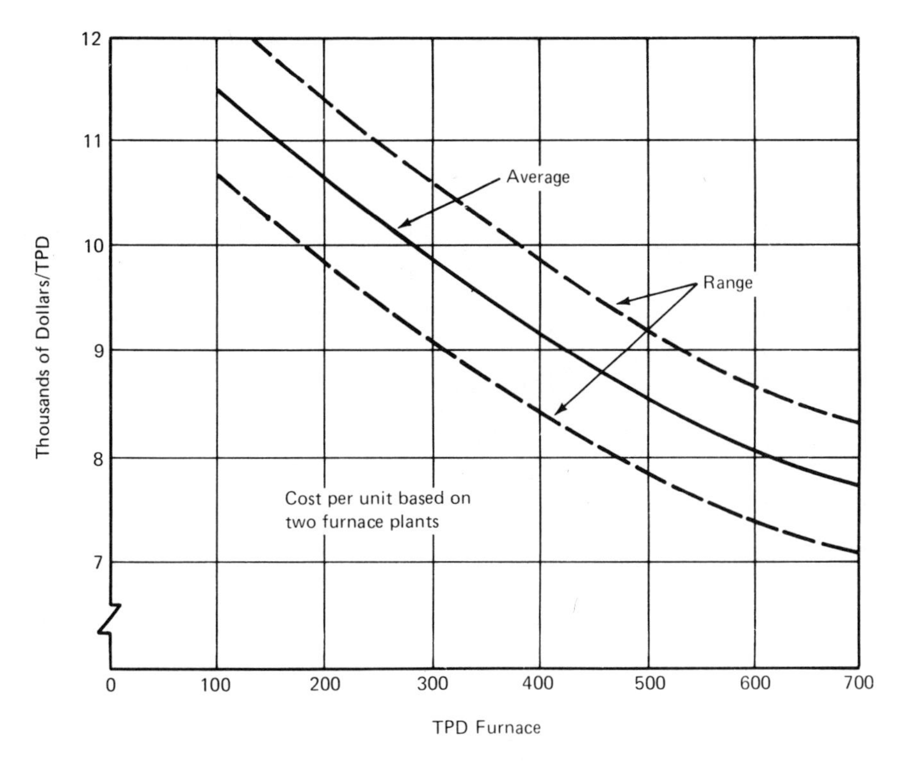

Figure 9-3. Estimated 1969 Cost of Incinerator Plant with Boilers and Precipitators. Source: Niessen *et al., Systems Study . . .*, p. VII-144.

the higher capital investment, is more costly. This illustrates the necessity for steam credits to justify water-wall operation.

An Accounting System

The lack of a consistent procedure for reporting operating costs is indicative of a general need throughout the industry for a standard accounting system. The U.S. Public Health Service has set up a model for such a system, which suggests a method using cost centers, forms, and reports.[1] In explaining the need for such a system, they make the following observations:

> Effective solid waste management requires an adequate information system including data on activity and the costs of operation and ownership. Although a cost accounting system represents only one

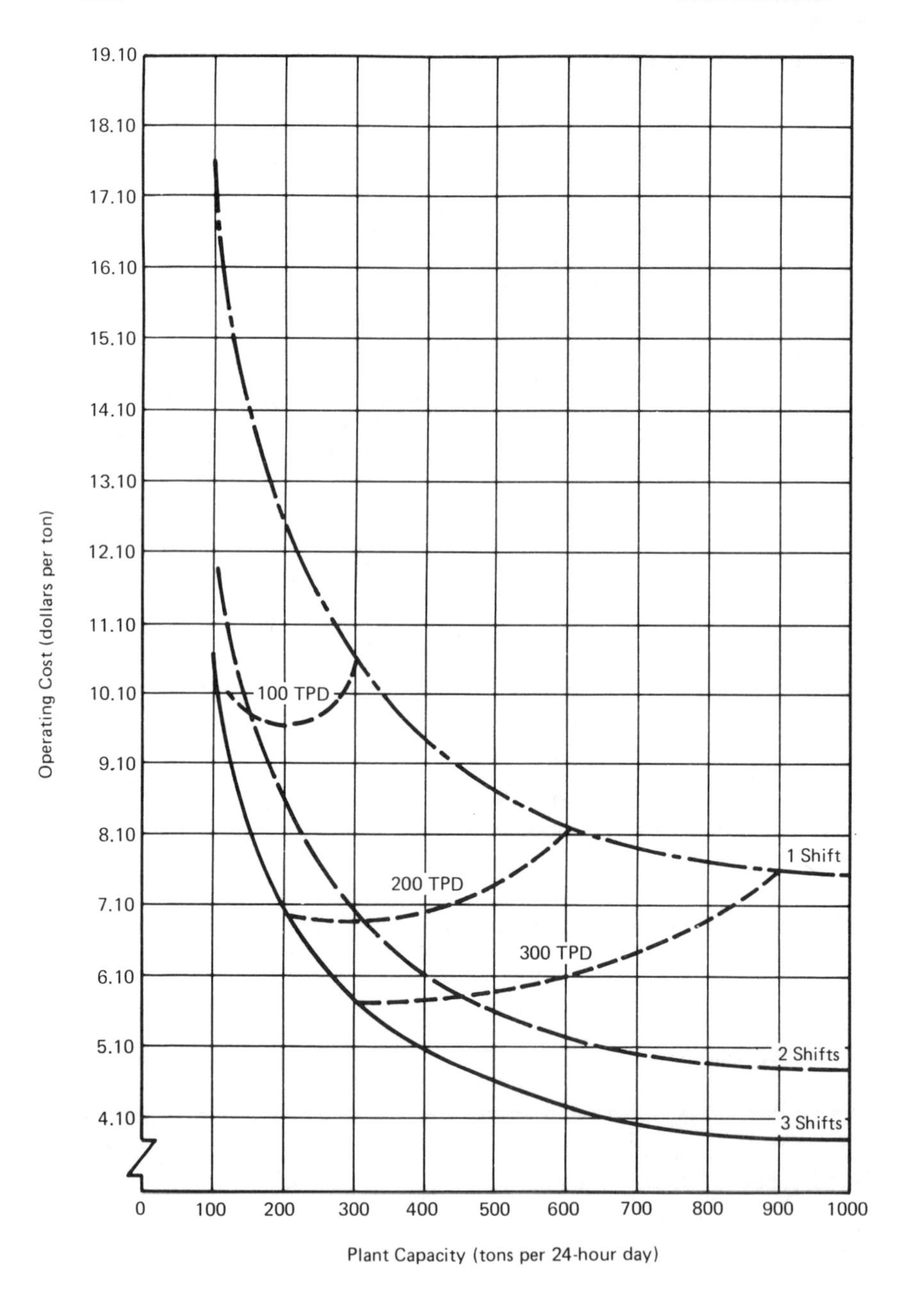

Figure 9-4. Total Operating Cost Versus Plant Capacity for Batch-Feed Units. Source: Niessen *et al., Systems Study . . .*, p. VII-102.

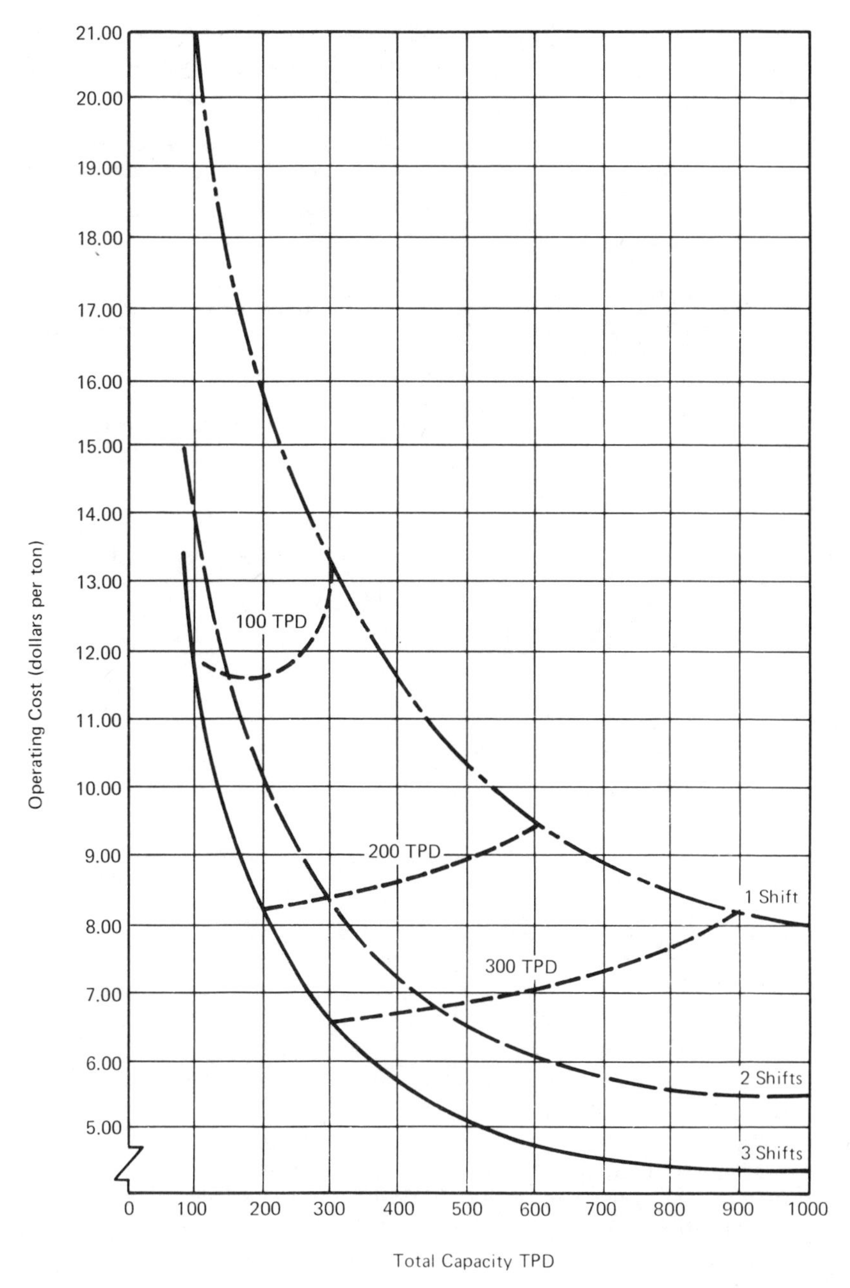

Figure 9-5. Total Operating Cost Versus Plant Capacity for Continuous-Feed, Rectangular-Construction Units. Source: Niessen *et al.*, *Systems Study . . .*, p. VII-108.

Table 9-1
Estimated Cost of Incineration Using Batch Incineration Systems (1969 Dollars)

	Reference location:		*Boston, Massachusetts, mid-1969*	
	Plant capacity:		*400 TPD (two, 200-TPD furnaces)*	
	Load factor:		*250 days per year; 3 shifts/day*	
	Annual consumption:		*100,000 tons refuse/year*	
	Capital investment:		*$2,300,000*	
	Units	*Dollars/unit*	*Units/ton refuse consumed*	*Dollars/ton refuse consumed*
Variable costs				
Electricity	KWH	$0.015	22.3	0.33
Water	M Gal	$0.30	0.90	0.27
				$0.60
Semi-variable costs				
Labor				
1 crane operator per shift @ $5.00/hour				0.31
2 furnace operators (stoking & dumping) per shift @ $4.00/hour				0.38
1 charging-floor man per shift @ $3.00/hour				0.19
2 residue-removal men per shift @ $3.00/hour				0.38
1 supervisor @ $14,000/year				0.14
1 weigh-scale operator @ $5,500/year				0.06
				$1.46
Overhead @ 40% of labor & supervision				0.58
Maintenance and repair @ 2.5% of capital investment annually				0.58
				$2.62
Fixed costs				
Amortization @ 4.7% of capital investment annually				1.08
Interest on loan @ 3.0% of capital investment annually				0.69
				$1.77
Total costs				$4.98

Source: Niessen *et al., Systems Study . . .* , p. VII-101.

part of the total system, it does facilitate the collection and later utilization of the data obtained.

Present information on incineration and its associated costs is both inadequate and non-standardized. The proposed system provides a guide to the type and quantity of information to be collected, its classification, and the method of collection. Incinerator supervisors and heads of agencies responsible for their operations will find the system useful.

A cost accounting system can aid a community on controlling the costs and performance of its incinerator operations, as well as aid in formulating future plans.[2]

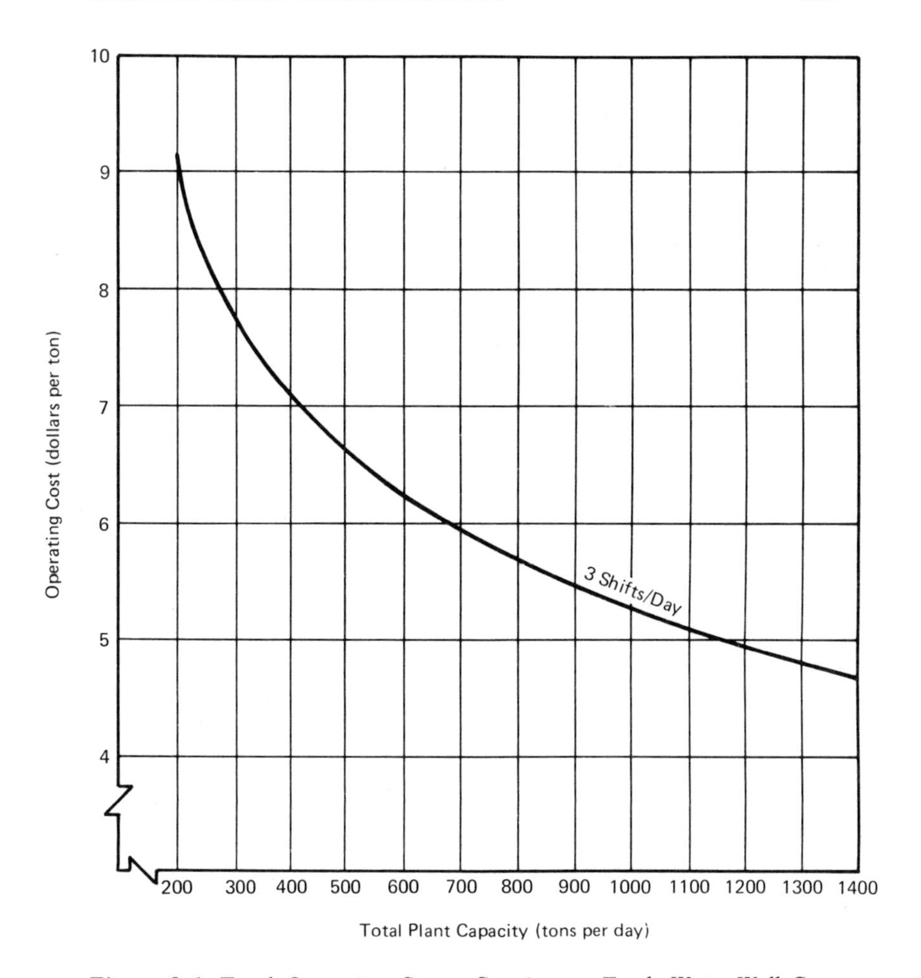

Figure 9-6. Total Operating Costs–Continuous-Feed, Water-Wall Construction Incinerator Systems (Grate-Burning, No Steam Credits). Source: Niessen *et al., Systems Study . . .*, p. VII-146.

The system sets up cost centers as follows: direct cost centers including Receiving and Storage, Volume Reduction, and Effluent Handling Treatment. Repairs and Maintenance is set up as an indirect cost center. These are said to be a minimum, and some larger units may require more detailed sub-divisions, or addition of other cost centers (salvage or heat utilization, for example). Figure 9–7 shows the flow of materials through the three direct cost centers, while Figure 9–8 shows how the system allocates operating costs to the various cost centers, and combines them with capital costs to arrive at a total annual cost.

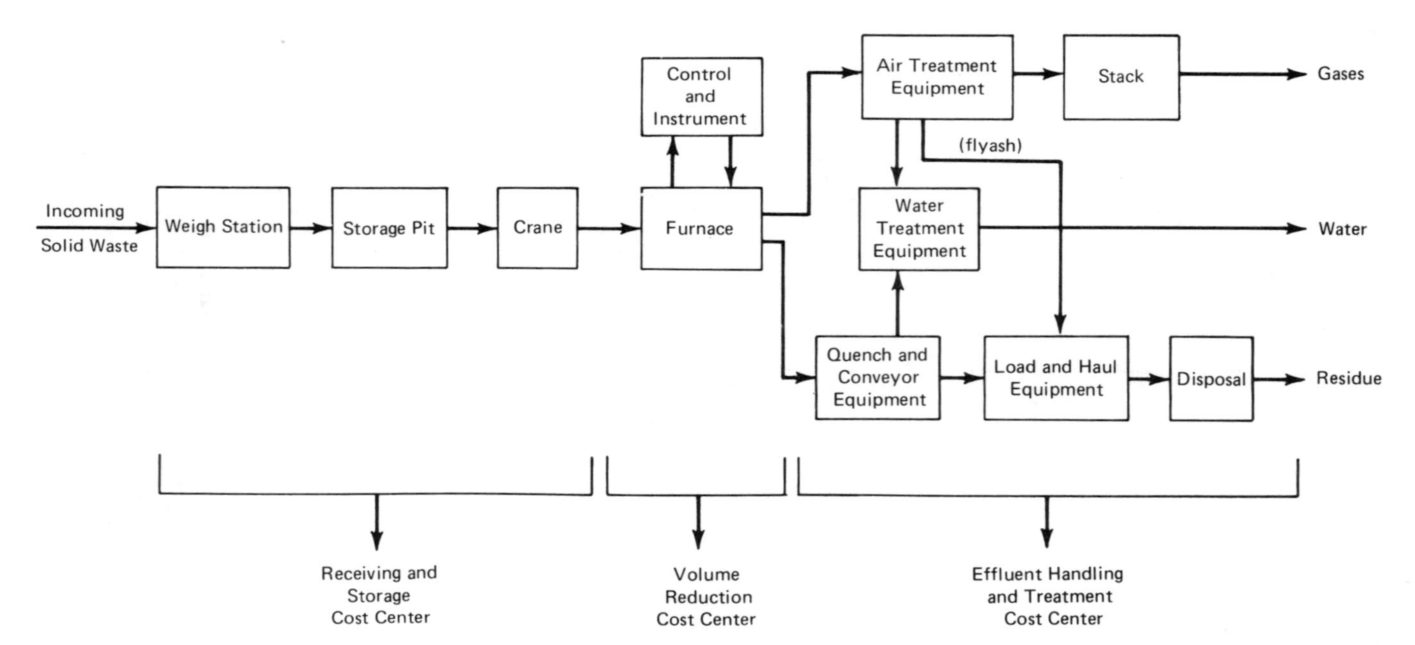

Figure 9-7. Incinerator Cost Centers. Source: Eric R. Zausner, *An Accounting System for Incinerator Operations*, p. 4.

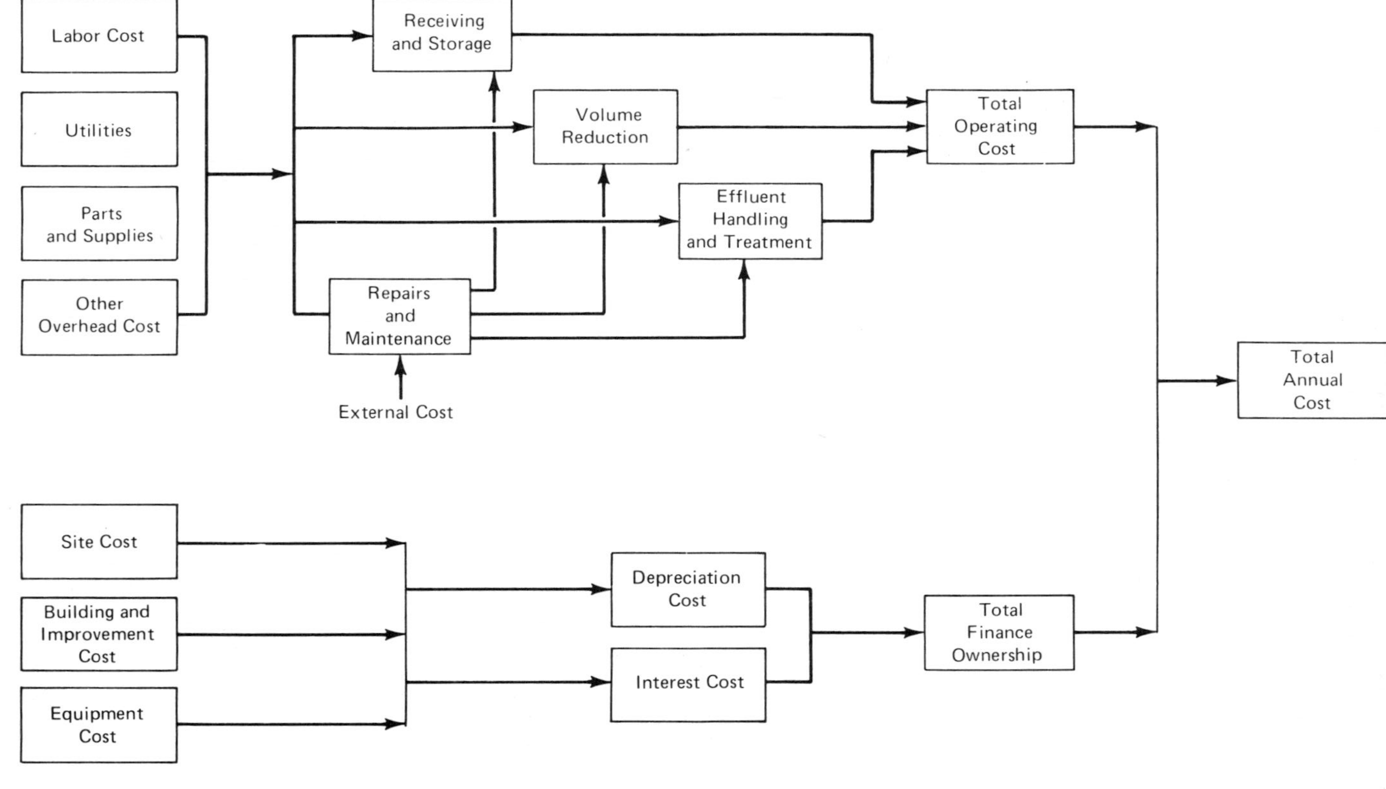

Figure 9-8. Allocation of Costs. Source: Eric R. Zausner, *An Accounting System for Incinerator Operations*, p. 5.

Six report forms are suggested: the Weekly Labor Report, the Daily Truck Record, the Daily Report on Incinerator Operations, the Incinerator Capital Investment Report, the Incinerator Operations Summary, and the Incinerator Total Cost Report. Sample forms are provided in the Public Health Service booklet.

10 New Developments

Several of the new techniques that have recently been applied to municipal refuse incineration have been, in some cases, only on an experimental or even theoretical level. Some involve new technology, while others represent merely a new application for older technology. This section describes four of the more novel developments—total incineration, fluid-bed incineration, pyrolysis, and open-pit incineration—along with the long-range outlook for their applications.

Total Incineration

The term "total incineration" denotes a process using extremely high temperatures (approximately 3000 degrees F.) to melt the entire residue to a slag, which is then drained from the furnace and solidified. These units are sometimes referred to as high-temperature or slagging incinerators. Further, if materials are to be removed from the residue, the effort makes use of the tendency for the molten fractions to segregate according to their different densities—that is, molten metal settles to a zone below the molten siliceous materials.[1]

The principal objectives of total incineration are:

1. Maximum volume reduction of solid waste (approximately 97.5 percent).
2. Complete combustion or oxidation of all combustible materials, producing a solidified slag that is sterile, free of putrescible matter, compact, dense, and strong.
3. Elimination of the necessity for a larger residue-disposal operation adjacent to the incinerator.
4. Complete oxidation of the gaseous products of incineration with discharge to the atmosphere after adequate treatment for air pollution control.[2]

There are several different designs of furnaces for total incineration. Figures 10-1 through 10-7 present schematic diagrams of seven such incinerator designs (from the Zinn, LaMantia, and Niessen article), while a summary comparison of these seven designs is presented in Table 10-1. All of these designs have in common the same advantages and disadvantages, problem areas and

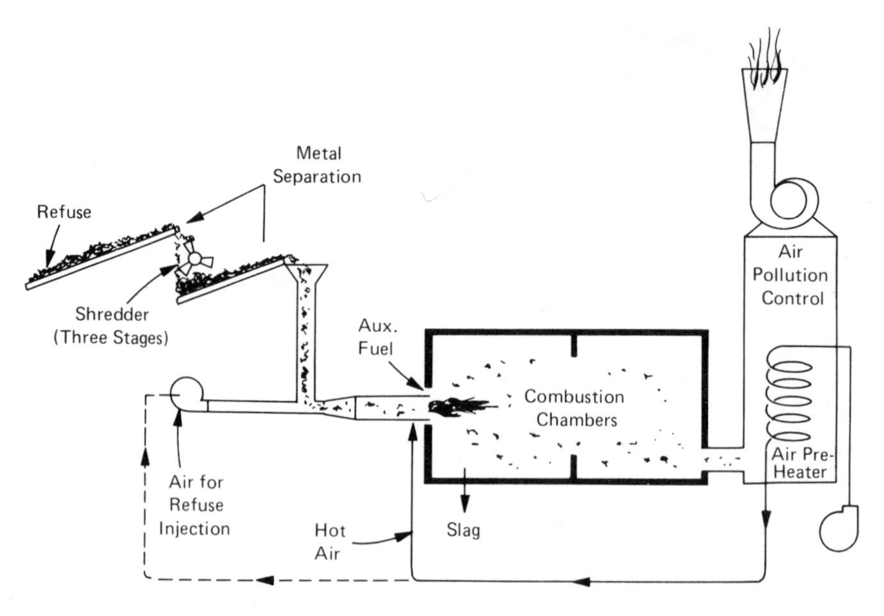

Figure 10-1. Sira System. Source: R.E. Zinn, C.R. LaMantia, and W.R. Niessen, "Total Incineration," *Proceedings of the National Incinerator Conference* (1970), ASME, p. 119.

economics; they are summarized from the Zinn and LaMantia article over the next paragraphs.

Advantages

Total incineration provides for the highest possible reduction in refuse volume. In addition, the residue is free of putrescible and combustible matter and is easily handled. High temperatures ensure complete oxidation of the odorous components of flue gases, and the smaller volume of combustion air used reduces the load on air pollution control equipment. (Most high-temperature incinerators operate with significantly lower excess air requirements than conventional incinerators.)

Disadvantages

The high temperatures involved in total incineration are likely to produce higher levels of fumes and nitric-oxide emissions. The equilibrium nitric-oxide

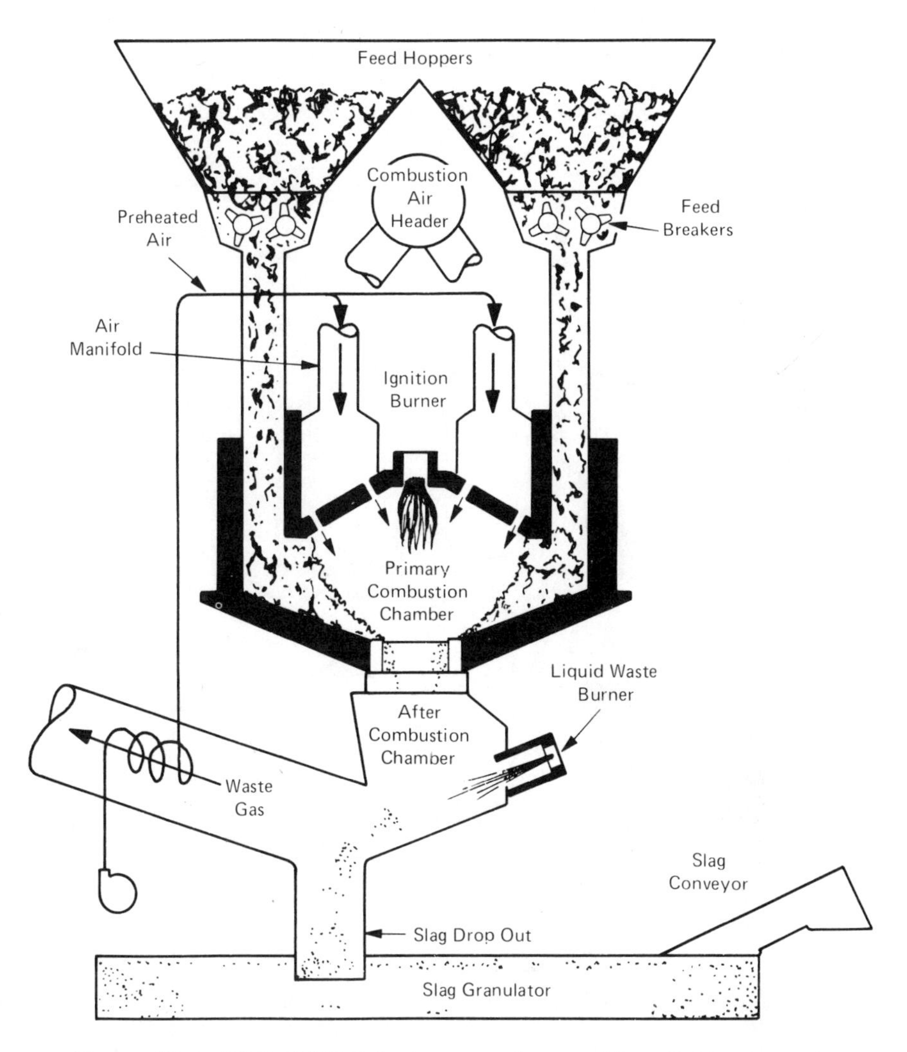

Figure 10-2. DRAVO/FLK Incinerator. Source: Zinn, LaMantia, and Niessen. "Total Incineration," ASME, p. 118.

concentration for combustion gases with 10 percent excess air is about 200 PPM at 2000 degrees F. and 2000 PPM at 3000 degrees F. Other pollutant levels will vary with the system design. Supplementary heat is also required, which increases operating costs. The overall process is more complex than conventional incineration, and increased safety precautions are required.

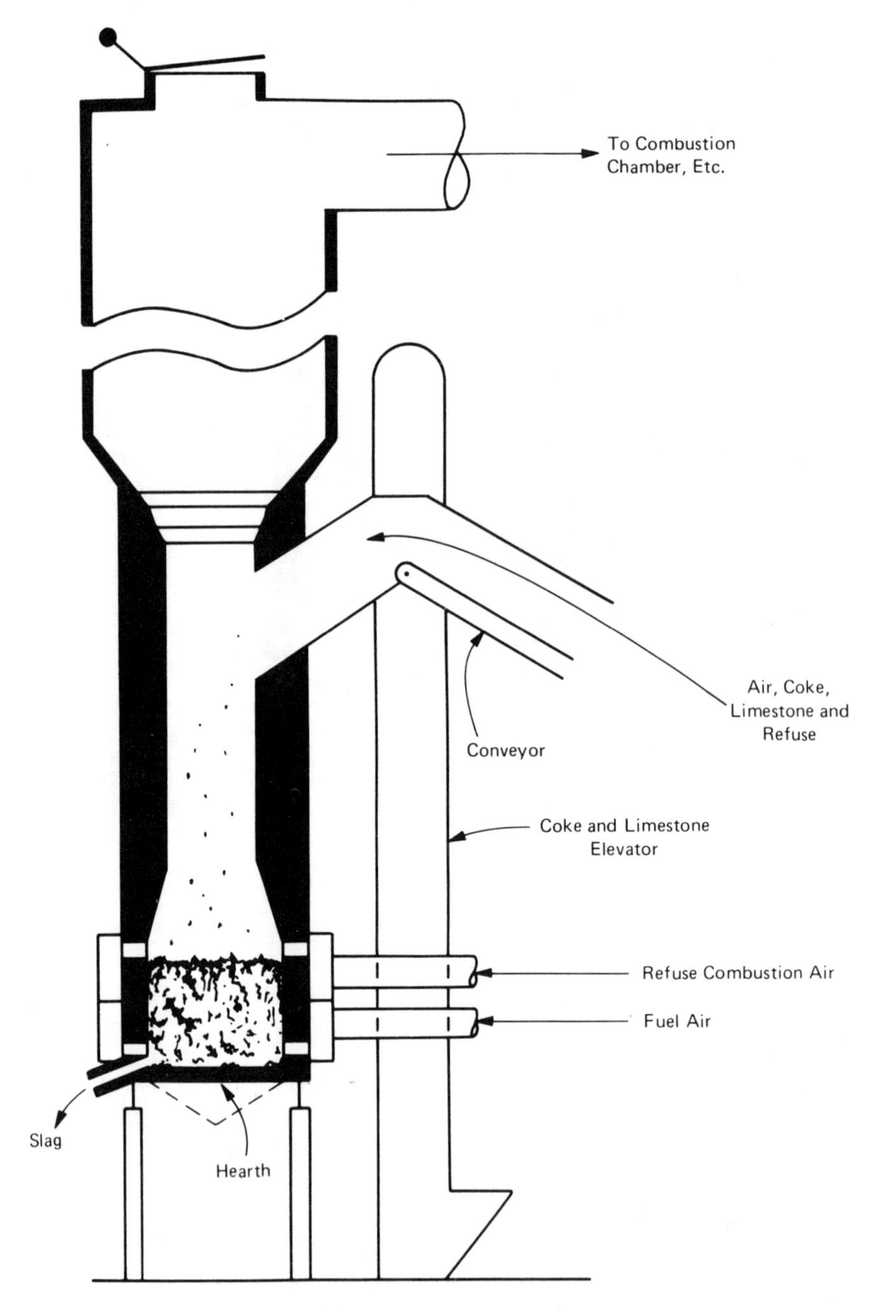

Figure 10-3. American Thermogen System. Source: Zinn, LaMantia, and Niessen, "Total Incineration," ASME, p. 118.

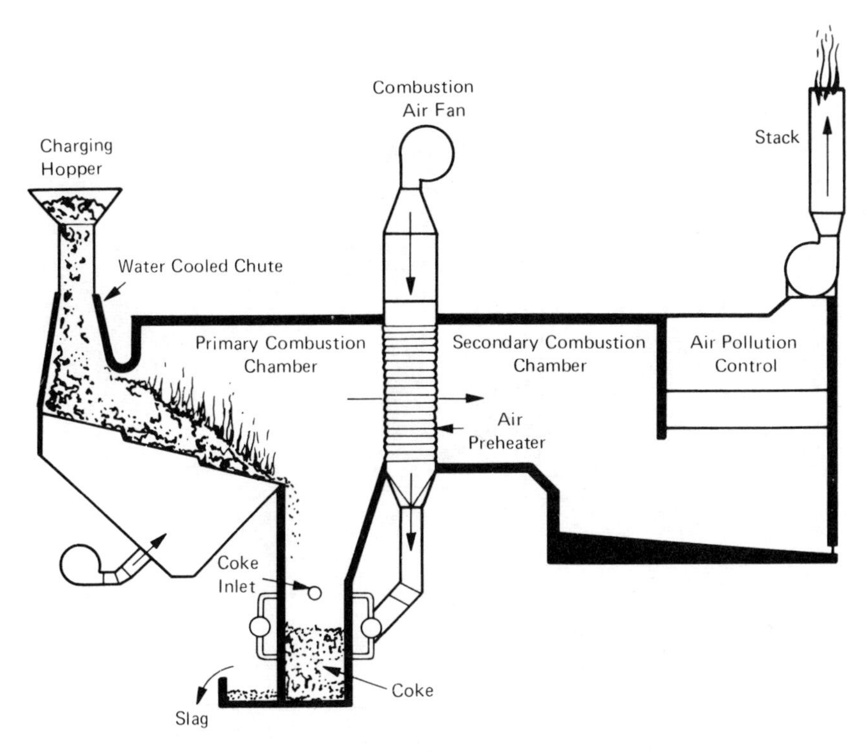

Figure 10-4. Ferro-Tech System. Source: Zinn, LaMantia, and Niessen, "Total Incineration," ASME, p. 120.

Problem Areas

A higher skill level for operators of high-temperature incinerators is required than for operators of conventional incinerators. Engineering problems involved with selection of materials, process equipment, instrumentation, and maintenance must be solved, due to the higher temperatures and operational sensitivity. Difficulties may also be encountered with the fluidity of the slag. Finally, as with any incinerator, continual attention must be paid to pollution control procedures.

Economics

It is likely that total capital costs for a total incineration plant will not be significantly different from those for conventional systems. Operating costs

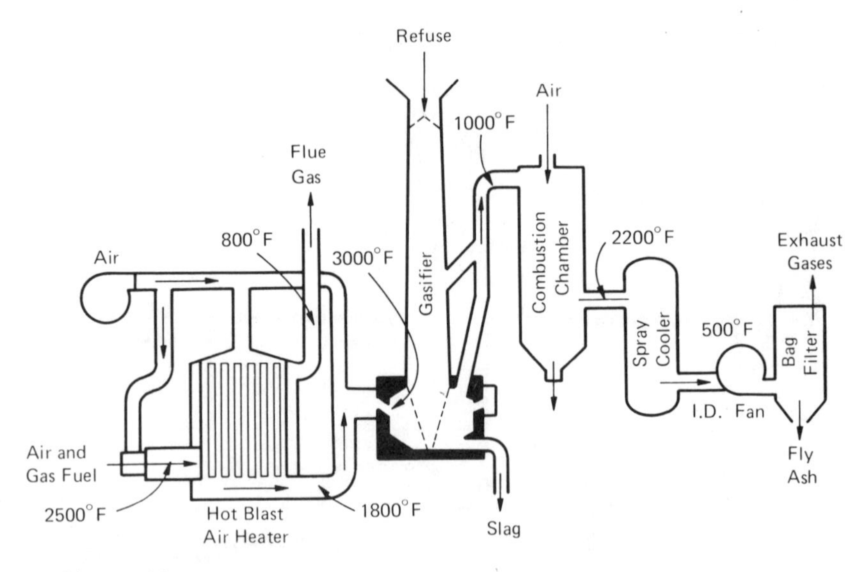

Figure 10-5. Torrax System. Source: Zinn, LaMantia, and Niessen, "Total Incineration," ASME, p. 120.

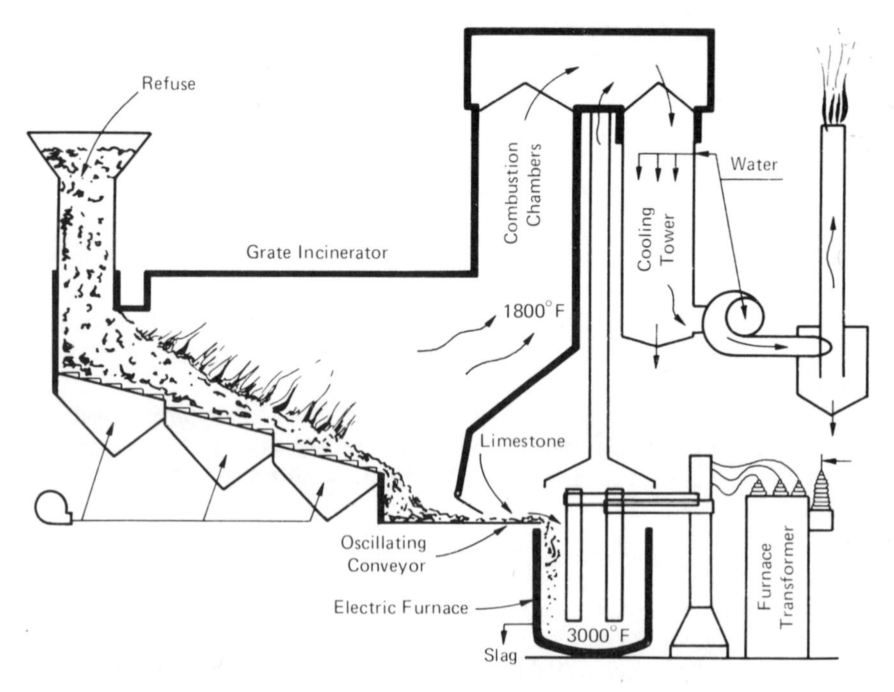

Figure 10-6. Electric-Furnace System. Source: Zinn, LaMantia, and Niessen, "Total Incineration," ASME, p. 121.

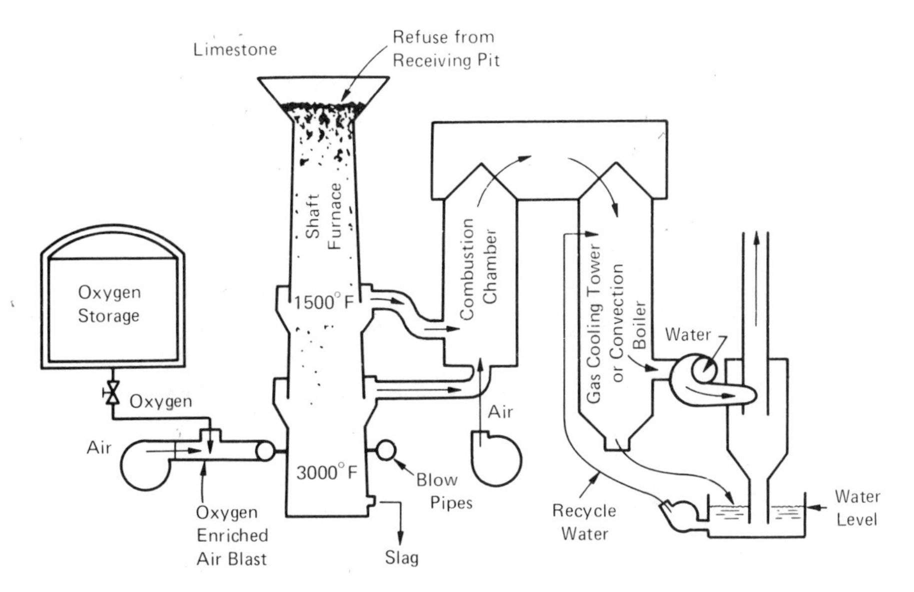

Figure 10-7. Oxygen-Enrichment System. Source: Zinn, LaMantia, and Niessen, "Total Incineration," ASME, p. 121.

will probably range from $1 to $2 more per ton of refuse processed, due primarily to auxiliary fuel requirements.

Suspension Firing

Suspension firing is not really a separate, complete process for incineration, but is rather a technique that may be incorporated into various incinerator designs. As the name implies, the refuse is burned largely while suspended in air. It is evident that this procedure requires shredding of the refuse prior to charging into the furnace.

Suspension firing is widely practiced in the burning of pulverized coal for power generation. Its application to solid waste incineration has been widely discussed. One suggested system for such a furnace is called "tangential firing," and its operation has been summarized from the Regan article as follows:

> Four pneumatic lines deliver refuse to each elevation of tangential nozzles, one line per corner. The refuse and the heated combustion air are directed tangentially to an imaginary cylinder in the center of the furnace. Fuel and air are mixed in a single fireball. This procedure precludes the possibility of poor distribution of fuel and air; it also permits operation with less excess air, thereby reducing the size of the flue gas

Table 10–1
Comparisons of Total Incineration Processes

	Dravo/ FLK	*American Thermogen*	*SIRA*	*Ferro-Tech*	*Torrax*	*Electric Furnace*	*Oxygen Enrichment*
Added capital cost (per installed ton) over conventional incineration	← None to $2000/ton depending on choice of auxiliary equipment →						
Operating cost (per ton refuse) over conventional incineration	← None to about $2.00/ton (mostly energy) →						*
Auxiliary energy required	Some gas or oil	Coke or gas	Some gas or oil	Coke	Gas	Electric power	Bulk oxygen
Air preheat from recuperator	Yes	Possible	Yes	Yes	Possible	No	No
Potential NOx air pollution	High	High	Medium	Medium	High	Lowest	High
Relative size of APC equipment	Medium	High	Medium	Medium	Medium	Low	Medium
Operating skill required	Medium to high	High	Medium to high	High	High	Medium to high	High
Shredding required	Yes	No	Yes		No	No	No

*Oxygen required is 0.3 to 0.4 ton/ton refuse, depending on moisture.

Source: Zinn, La Mantia, and Niessen, "Total Incineration," ASME, p. 122.

> cleaning equipment. The refuse nozzles can be tilted upward or downward to accommodate variations in refuse characteristics and load. With tangential firing, the fuel particles have a longer residence time in the hottest furnace zones, thereby assuring complete combustion of waste fuels with low heat content.
>
> As the burning refuse particles spin downward, additional preheated combustion air is introduced in the lower furnace through multiple rows of tangential nozzles. This continues the combustion process and maintains particle momentum. Since the larger refuse particles will not be completely burned in suspension, a small grate may be required in the bottom of the furnace to complete combustion of the larger particles and to remove ash.
>
> Oil or gas firing is usually included for use during start-up and as a secondary fuel.[3]

A tangentially fired incinerator has been operating at the Eastman Kodak plant in Rochester for about two years. The unit was designed to burn 180 tons per day of industrial refuse and 114 tons per day of sludge.[4] Initial reports of the performance of the incinerator indicate that the incinerator itself is quite successful although some material handling problems with the shredded refuse before it enters the incinerator were encountered. Another all-shredded-refuse-fired incinerator is scheduled to become operational at a General Motors plant in Pontiac, Michigan, in early 1973. As previously explained, the only all-shredded-municipal-refuse-fired incinerator became operational at Hamilton, Ontario, in 1972.

In addition to the large incinerators cited here, work is also being done to develop smaller vortex incinerators. "Development of Vortex Incinerator with Continuous Feed," by C.H. Schwartz, A.A. Orning, R.B. Snedden, J.J. Demeter, and D. Bienstock, in the *Proceedings of the National Incinerator Conference* (1972), discusses one such experiment. The Bureau of Mines is studying an incinerator where the refuse is burned in a vortex suspension. "Operating Experience in the Suspension Burning of Waste Materials in Cyclone Incinerators," by R.G. Mills and L.G. Desmon, also from the 1972 Conference, presents research being done on cyclone incinerators, which are quite similar to the vortex types.

Fluid-Bed Incineration

Fluidized beds have long found application in the chemical industry for certain reactions between solids and gases where considerable heat is evolved. Basically, the fluidized-bed reactor consists of a cylindrical tube. Sand is used to form the bed inside the tube, and it rests on a porous plate, through which a gas (air, in the case of incineration) may be forced. As the gas pressure increases, the sand becomes agitated and assumes fluid characteristics. When the bed is heated, an essentially constant temperature gradient is maintained throughout the bed, and the rate of heat transfer is very rapid due to the violent movement of the solid wastes. This high rate of heat transfer results in rapid ignition and combustion. The heat of combustion is in turn rapidly absorbed by the bed. This process has been described as a "thermal flywheel"—that is, it gives up and receives energy.[5] Heat-transfer surfaces can maintain a constant temperature in the bed.

Figure 10-8 is a basic fluidized-bed reactor, while Figure 10-9 shows how it might be adapted for solid waste incineration.

For fluidized beds in the incineration of municipal solid wastes, the relatively low temperature and simplicity of construction may result in fairly low capital investments, and the heat-transfer characteristics seem to make it attractive for heat recovery. With the use of auxiliary fuel, temperature could be kept constant and power generation rates could be more easily controlled than with conventional furnaces. "Potential Advantages of Incineration in Fluidized Beds," by R.C. Bailie, D.M. Donner, and A.F. Galli, in the *Proceedings of the National Incinerator Conference* (1968), further discusses the positive aspects and processes of fluidized beds.

Combustion Power, Inc. is working on the development of a system for municipal solid waste management, incorporating fluidized-bed incineration with heat recovery to drive a turbine that in turn creates electricity. In the system shown in Figure 10-10, refuse is shredded into particles approximately two inches in size and dried and classified. The combustible portion of the refuse is transferred to a storage unit that meters the shredded waste. Passing through the air lock, the refuse is mixed with 600 degrees F. air as it enters the fluid-bed combustion chamber. The high heat-transfer rates promote rapid and complete combustion of the refuse. The bed is expected to operate at about 1650 degrees F. The entrapped particles in the hot exhaust gases are removed by a two-stage, inertial separator before the gas is used to drive the turbine. The turbine exhaust gas, which is approximately 930 degrees F. and at atmospheric pressure, can be passed through an optional waste-heat boiler.

A typical plant as conceived by Combustion Power could consume 400 tons of solid waste per day and produce approximately 9000 kilowatts of electrical power.

Nevertheless, fluidized beds as a concept still harbor a number of unsolved potential problems. Among them are the following:

1. Fluidized beds are not generally operated with the range of particle sizes expected for the Combustion Power unit, where relatively fine sand particles are admixed with much coarser refuse and fine ash particles. Air distribution will be a problem in units of commercial scale.

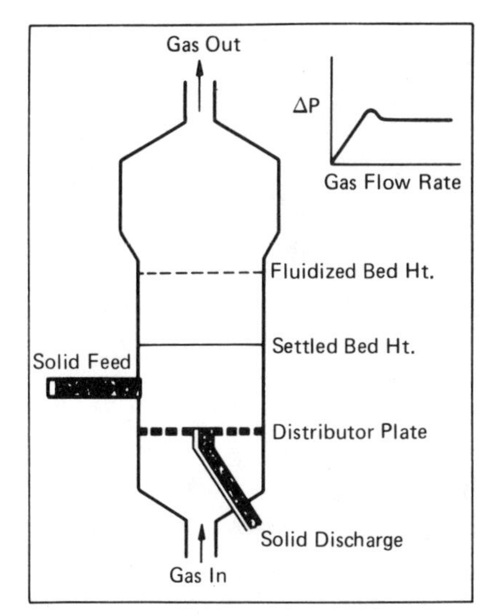

Fluidized bed reactor. Gas flow through the porous plate can be controlled to any desired rate; at low flow rates bed remains in its original "packed" state, and pressure drop across bed increases with flow rate until it is equal to the downward force exerted by the solids resting on the porous plate. Bed begins to expand at this point (incipient fluidization), allowing more gas to pass through bed at same pressure drop.

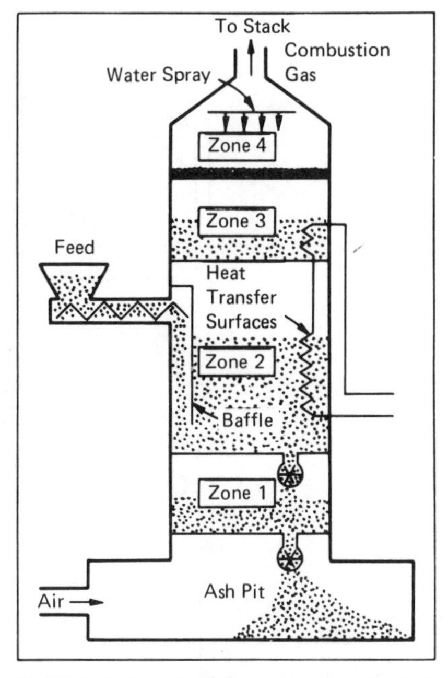

Fluidized bed for the clean combustion of solid wastes and heat recovery. Incoming air is preheated in Zone 1 (fluidized ash bed) before it enters primary combustion chamber or fluidized sand bed Zone 2. In the secondary fluidized sand bed, Zone 3, combustion is completed and gases cooled; in flooded packed bed, Zone 4, the gases are further cooled and scrubbed.

Figure 10-8. Basic Fluidized Bed Reactor. Source: R.C. Bailie, "Solid Waste Incineration in Fluidized Beds," *Industrial Water Engineering*, November 1970, pp. 22-25.

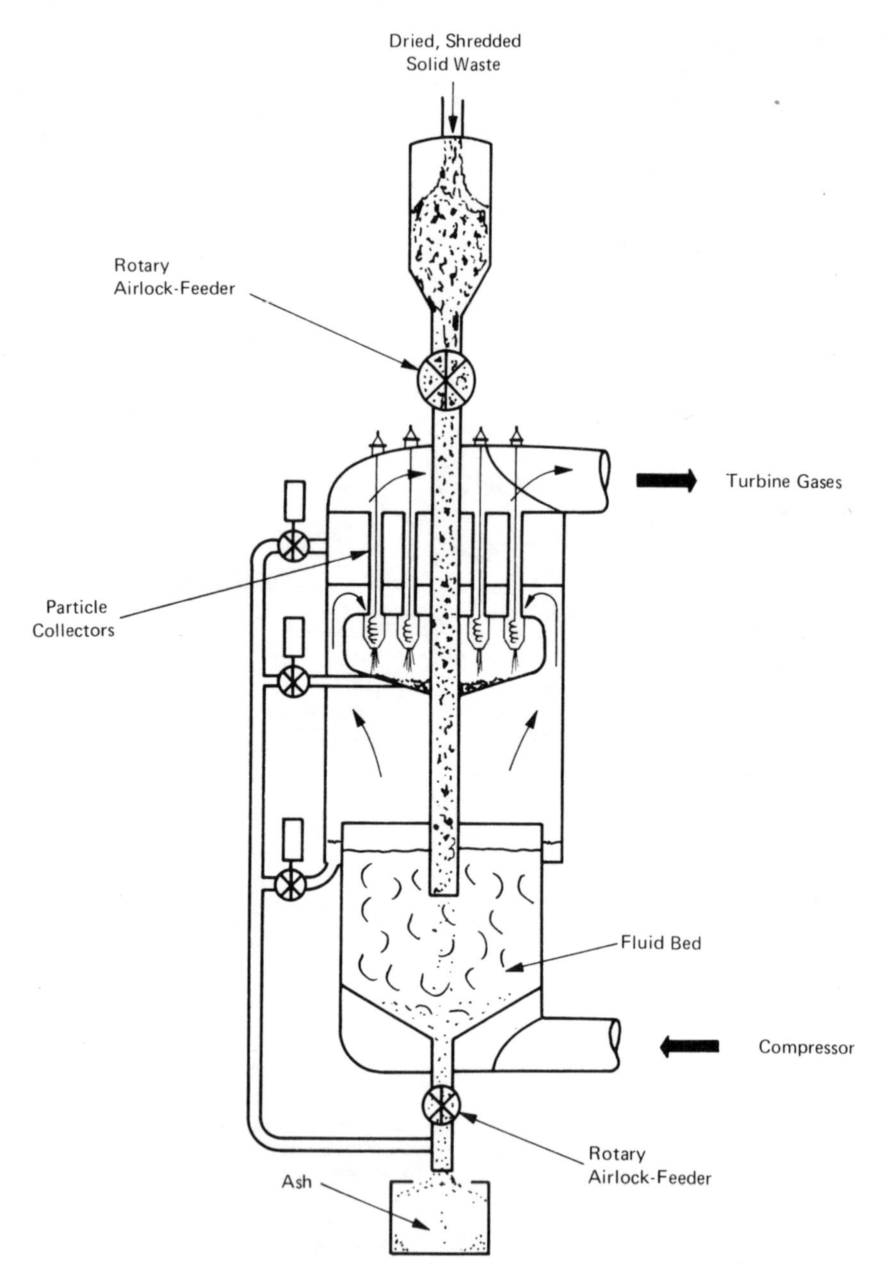

Figure 10-9. Fluid-Bed Reactor Adapted for Solid Waste Incineration. Source: Combustion Power Company.

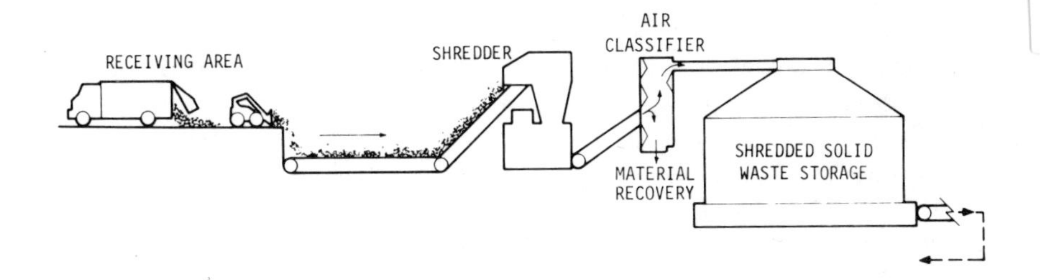

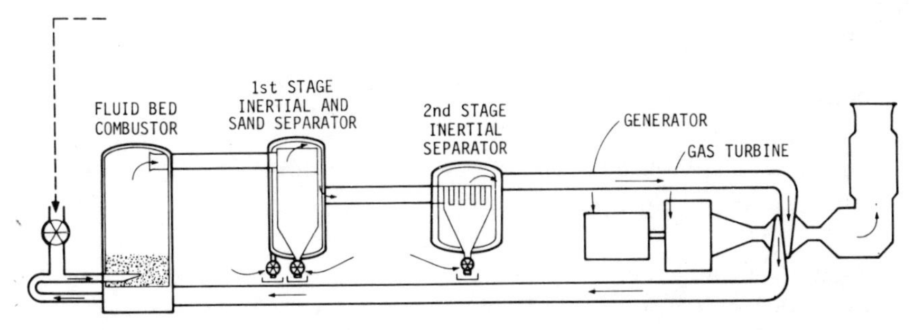

Figure 10-10. Combustion Power, Menlo Park System Flow Chart. Source: Combustion Power Company.

2. Defluidization of the bed may be experienced due to particulate agglomeration fluxed by melted particles of inorganics.
3. The operation of both rotary air lock feeders seems extremely uncertain, particularly the coarse particle withdrawal valve.
4. Demands on particle collectors are extremely severe for trouble-free turbine operation. Operation at 1500 to 1600 degrees F. is virtually unknown in past practice.

The CPU-400 has been operated using 100 percent refuse as a fuel but the operation has not yet been extensive enough to test the life of the turbine blades. Combustion Power is presently preparing for a continuous 48-hour all-refuse-fired test.

Pyrolysis

Strictly speaking, pyrolysis is not a form of incineration, but a type of chemical processing. It is sometimes considered, however, in the same context as incineration when it is applied to solid waste processing. The refuse is not burned—i.e., oxidized—in pyrolysis, but rather is thermally converted in the absence of oxygen. Pyrolysis, or carbonization, has been used for many years for production of coke from soft coal. It is considered technically feasible to apply the same process to solid waste, thereby obtaining products such as gas, water, organic liquids, and char, while reducing the volume of the refuse.

The nature of the products produced by pyrolysis of various types of solid waste at 1500 degress F. is shown in Table 10-2. Table 10-3 gives the results of pyrolysis upon changing a "typical" mixture of organic refuse components containing 20 percent moisture and about 2 to 3 percent ash. Table 10-4 presents yields of gaseous products using "typical" refuse, and Table 10-5 presents the heating value of char obtained from the same tests.

Bureau of Mines

The Bureau of Mines is developing a pyrolysis unit for resource recovery; its work in this area began as early as 1929 with the contruction of a research

Table 10-2
Pyrolysis Yields from Various Refuse Components (At 1500 Degrees F.)

	Gas	*Water*	*CnHmOx*	*Char C + S*	*Ash*
Rubber	17.29	3.91	42.45	27.50	8.85
White Pine Sawdust	20.41	32.78	24.50	22.17	0.14
Balsam Spruce	29.98	21.03	28.61	17.31	3.07
Hardwood Leaf Mixture	22.29	31.87	12.27	29.75	3.82
Newspaper	25.82	33.92	10.15	28.68	1.43
Corrugated Box Paper	26.32	35.93	5.79	26.90	5.06
Brown Paper	20.89	43.10	2.88	32.12	1.01
Magazine Paper	19.53	25.94	10.84	21.22	22.47
Lawn Grass	26.15	24.73	11.46	31.47	6.19
Citrus Fruit Waste	31.21	29.99	17.50	18.12	3.18
Vegetable Food Waste	27.55	27.15	20.24	20.17	4.89

Source: E.R. Kaiser, "The Pyrolysis of Refuse Components," presented at the 60th Annual Meeting, AIChE, Nov. 26-30, 1967.

Table 10-3
Pyrolysis Yields at Various Temperatures (Typical Refuse)

Temp., °F	*Gases*	*Pyroligneous Acids and Tars*	*Char*	*Mass Accounted For*
900	12.33	61.08	24.71	98.12
1200	18.64	59.18	21.80	99.62
1500	23.69	59.67	17.24	100.59
1700	24.36	58.70	17.67	100.73

Source: Donald A. Hoffman and Richard A. Fitz, "Batch Retort Pyrolysis of Solid Municipal Wastes," *Environmental Science & Technology*, Vol. 2, Number 11, November 1968, p. 1025. Reprinted with permission. Copyright by the American Chemical Society.

Table 10–4
Yields of Gaseous Products from Pyrolysis (Typical Refuse)

Temp., °F	*900*	*1200*	*1500*	*1800*
SCF/lb	1.90	2.78	3.62	3.39
BTU/SCF	300	376	344	351
BTU/lb	569	1045	1245	1190
Mole percent				
H_2	5.56	16.58	28.55	32.48
CH_4	12.43	15.91	13.73	10.45
CO	33.50	30.49	34.12	35.25
CO_2	44.77	31.78	20.59	18.31
C_2H_4	0.45	2.18	2.24	2.43
C_2H_6	3.03	3.06	0.77	1.07

Source: Hoffman and Fitz, "Batch Retort . . . ,"

Table 10–5
Analysis and Heating Value of Char from Pyrolysis (Typical Refuse)

	Temperature, °F.			
	900	*1200*	*1500*	*1700*
Volatile matter, %	21.81	15.05	8.13	8.30
Fixed carbon, %	70.48	70.67	79.05	77.23
Ash, %	7.71	14.28	12.82	14.47
BTU per lb.	12,120	12,280	11,540	11,400

Source: Hoffman and Fitz, "Batch Retort . . . ,"

unit for coal. Its primary interest now is to develop a commercial package for pyrolysis, with emphasis upon recovery of useful products. The basic concept of this system is illustrated in Figure 10–11.

As previously suggested, the economics of the Bureau of Mines system depends upon the recovery of fuels. It is estimated that commercial operations, with credit for recovery of fuels, could operate at costs ranging from $6.00 per ton for a 500-ton-per-day unit, to $2.00 or less per ton for a 2,500-ton-per-day unit.[6]

Other Units

Not all the work in pyrolysis is concerned with production of gas, oil, or other products. Monsanto Company has conducted pilot tests with the goal of using the principles of pyrolysis in what is termed primarily a waste disposal system. This system is illustrated in Figure 10–12.

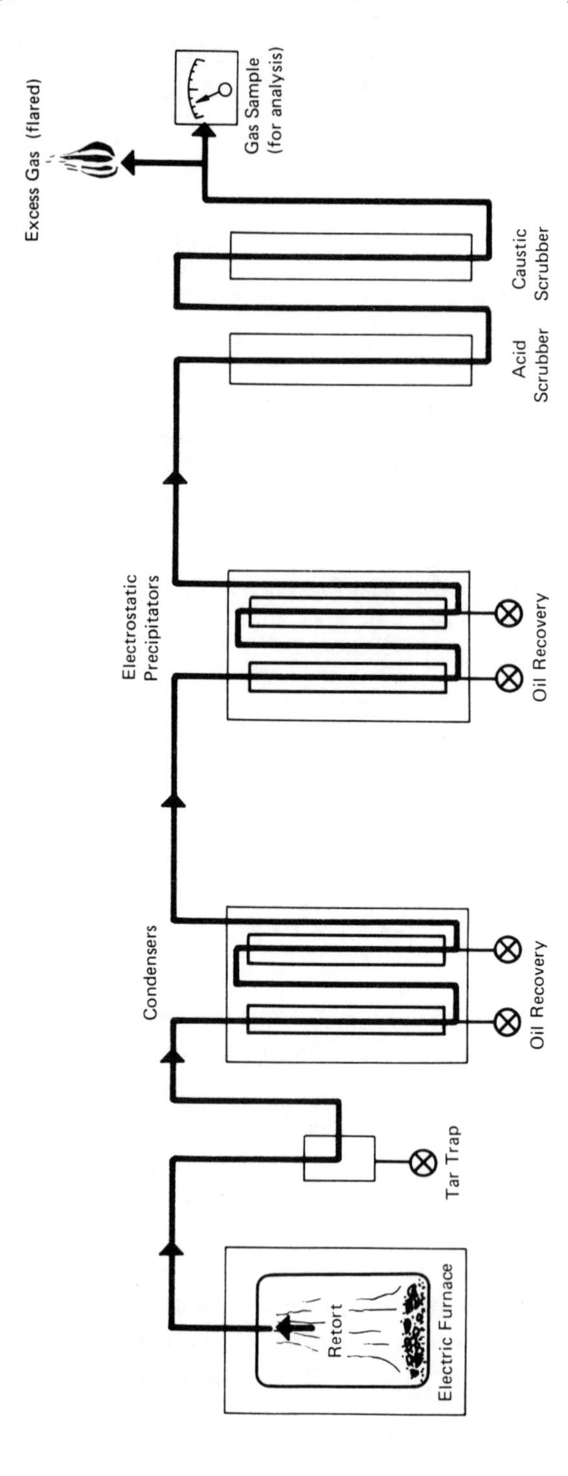

Figure 10-11. Bureau of Mines Pyrolysis Process. Source: "Pyrolysis of Refuse Gains Ground," *Environmental Science & Technology*, Vol. 5 (April 1971), p. 310. Reprinted with permission. Copyright by the American Chemical Society.

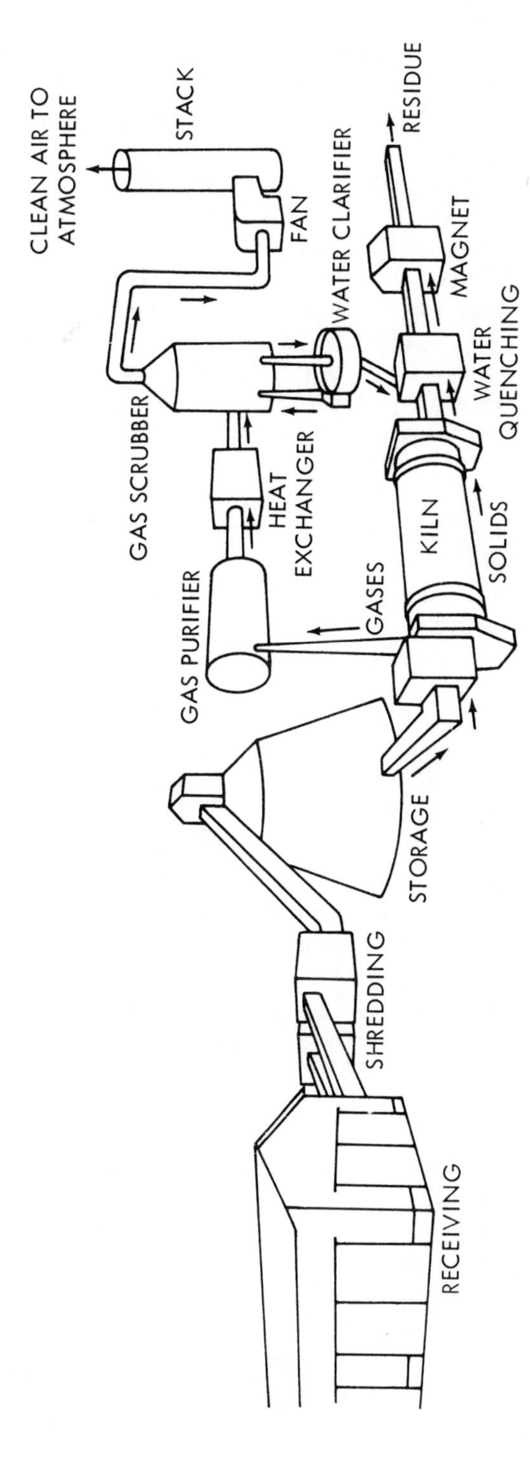

Figure 10-12. Monsanto Landgard System. Source: Monsanto Enviro-Chem Systems, Inc.

Shredded refuse is fed continuously to a refractory-lined, inclined, rotary kiln. The refuse is heated by burning natural gas inside the kiln, which distills moisture and combustible gases and vapors from the refuse. A limited amount of air is admitted to the kiln. Residue discharged from the kiln through a water seal contains char, metal, glass, and other inert materials. The kiln gases are burned to completion in a secondary combustion chamber, after which the gases are cooled, cleaned, and discharged to the atmosphere.

The Monsanto Landgard unit, it is reported, would have total costs of roughly two-thirds of those of pollution-free incinerators per ton without allowing for resource recovery. Since Monsanto feels that the most urgent need is for disposal, it has designed its Landgard system so as not to be economically dependent on resource recovery. It is stated, however, that the system will be flexible enough to allow for such recovery, if it proves economically advantageous to do so. Construction is underway on a 1000-ton-per-day plant in Baltimore. The plant, partially funded by EPA, is expected to be completed by 1975. Much of the solid waste will be converted into fuel gas and the gas, in turn, burned to produce steam.

Garrett Research and Development Corporation also has designed a pyrolysis plant and will build a 200-ton-per-day facility in San Diego County, California, in 1974 with an EPA grant. The Garrett process requires primary shredding to approximately two inches, air classification, drying, secondary shredding and fine screening to prepare the refuse for pyrolysis. Ferrous metals are removed magnetically, and the glass and non-ferrous components are ground in a rod-mill and screened to separate a fine glass fraction. Pyrolytic oil and char are the main products yielded.

In addition, the Linde Division of Union Carbide Corporation has built a 200-ton-per-day plant in Charleston, West Virginia. Their system involves feeding waste into a vertical tube furnace, followed by ignition of the refuse and injection of a small amount of oxygen. The pyrolysis occurs toward the top of the furnace, but pyrolyzed refuse eventually drops to the lower combustion zone and the off-gases are taken to a cleaning system. The inorganic fraction of refuse becomes a molten slag which is quenched with water to form solid residue, and the cleaned gases become a low-sulfur fuel suitable for industrial use.

Open-Pit Incineration

Essentially, open-pit incinerators avoid the problem of high-heat flux by eliminating the furnace enclosure (see Figure 10–13).

The open-pit incinerator was originally developed at DuPont for the safe destruction of nitrocellulose that presents an explosion hazard in a conventional closed incinerator. The incinerator shown has an open top and an array of

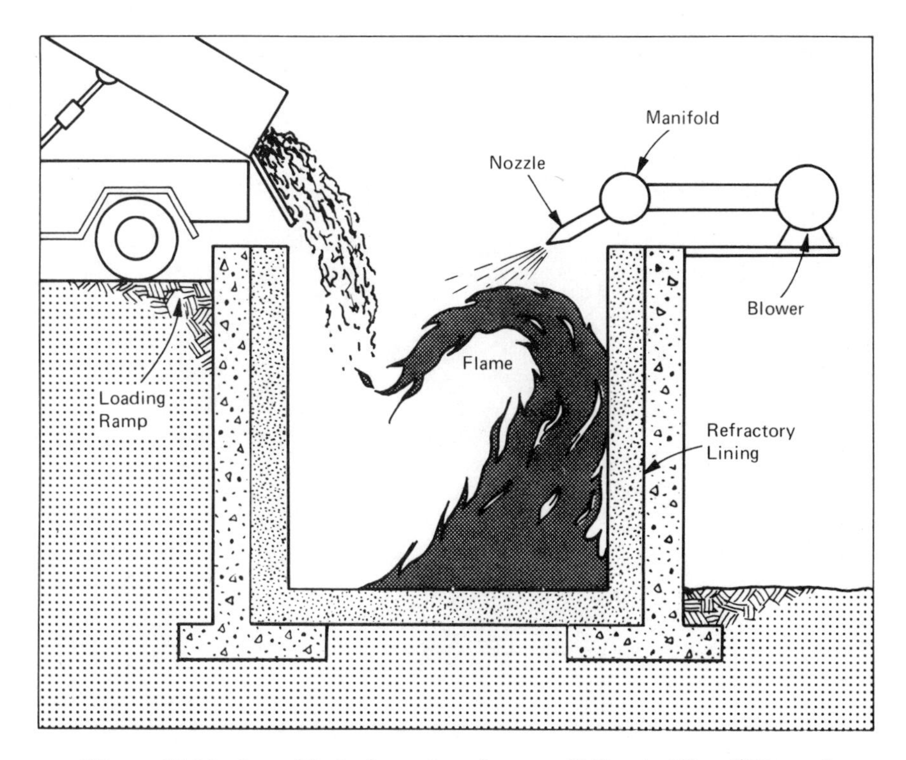

Figure 10-13. Open-Pit Incineration. Source: Philip A. Witt, "Disposal of Solid Wastes," *Chemical Engineering*, October 4, 1971, p. 67.

closely spaced nozzles that create a rolling action of high-velocity air over the burning zone. Very high burning rates, long residence times leading to complete combustion, and high flame temperatures are achieved. Visible smoke is readily eliminated and smuts are contained by screening. Normally, a screened enclosure is placed over the pit to contain large airborne particles and for insect and rodent control when not burning garbage.

The over-fire air is supplied from a manifold running along one edge of the pit, with alternative two- and three-inch nozzles directed downward at an angle of 25 to 35 degrees across the incinerator. Charging is from the opposite side of the nozzles from a loading ramp. The pit should be oriented so that the loading ramp is located upwind. The high-velocity air jets create turbulence in the burning zone, and the excess air aids complete combustion.

When the equipment is properly operated, the air pattern creates a sheet of flame under the air manifold on the back wall, rolling the flame across the top of the pit. Most of the particulates and unburned gases are returned to the burning zone—which tends to eliminate smoke.

11 On-Site Incineration

Another recent trend in dealing with municipal incineration is toward smaller systems located at the points where the refuse is generated. While a majority of these newer on-site systems are well-designed and effective, others are in need of updating or replacement. As with any incineration system, on-site facilities require intelligent supervision to prevent improper or incomplete combustion. In addition, it must be pointed out that on-site incineration severely hampers resource recovery.

Nevertheless, there are many advantages to on-site incineration:

1. Regular collection is not required (or may not be available), and hauling costs are saved. In many cases, the problem of storing trash until collection can be made is a formidable one. For example, large apartment complexes will generate quantities of refuse on a daily basis. Unless collection is also performed on a daily basis, a physical site must be maintained for storage. Storage areas take space that could be used for parking, recreation, or more apartments. Unless they are continuously maintained, these trash storage areas become a social blight as well as a health hazard.
2. Some refuse is best disposed of on-site. In some industrial and commercial operations, the waste generated is not acceptable to municipal collection systems and could create hazards in transportation and final disposal. A manufacturer may produce highly volatile waste such as plastics, paint thinners, or low-flash-point synthetics that can best be disposed of in a carefully controlled and supervised incineration process. Other industrial combustible wastes may be oversized. Classified or confidential documents are also candidates for on-site disposal. Some outputs from chemical plants, drug companies, food processors or handlers, biological research laboratories, hospitals, or other specialized businesses may best be incinerated under rigidly controlled observation.
3. On-site incineration reduces the cost and size of municipal collection and disposal facilities. Large commercial, industrial or multi-dwelling complexes producing vast quantities of refuse in a small area require many pick-ups, and consequently create traffic, noise and cost problems. If local governments enact legislation requiring that such businesses pre-treat their refuse by incineration, the cost to the city collection system would be reduced.

Criteria for the design of commercial, industrial, and domestic on-site incinerators will obviously differ from those for centralized municipal plants. This is due in part to the fact that on-site incinerators are required to handle much smaller quantities of refuse. While some municipal incinerators operate on a 24-hour basis with a constant supply of trash, the on-site facility may quite probably be used most effectively on a fractional-day basis. Thus, the system is subject to frequent start-ups and cyclic heating and cooling of the furnace that stress the various elements of the incinerator. And since the operator of the on-site incinerator may have other duties to perform, his training in proper use of the equipment is probably minimal.

Two general types of incinerators are available for on-site destruction of solid waste in compliance with codes, standards and air pollution regulations. They are available in standard sizes and capacities up to about 2000 pounds an hour, above which they are usually custom-designed. These incinerators are widely used in apartment buildings, commercial establishments, hospitals, and factories.

The Incinerator Institute of America (IIA) has issued standards for dimensioning and rating incinerators; the standards cover primary and secondary chambers, auxiliary fuel burners, air and temperature control, and flue gas cleaning.[1] Compliance with the standards is indicated by the IIA seal on the unit. The Institute has also published testing procedures and other technical information of interest to on-site incinerators.

Conventional Multiple-Chamber Incinerator

The hand-fired rectangular incinerator with cast-iron grates and firebrick construction has been improved during the last 15 years to assure compliance with restrictive air pollution regulations. Today this form of on-site incineration is a system of primary and secondary chambers, flame downpass, overfire air blower and nozzles, primary and secondary burners, temperature controller, draft control and flue gas washer.

Implementation of the basic design varies among the manufacturers within IIA specifications but Figure 11-1 illustrates the essential features.[2]

Warm-up of the chambers by the auxiliary burners is usually advisable before charging moist refuse. Temperatures over 1300 degrees F. are necessary to prevent smoke emission when hydrocarbons are present in the initial charges. The secondary chamber has the function of settling out the larger particles entrained in the gases leaving the primary chamber and burning them. Also, the burning of smoke and combustible gases is finished in the secondary chamber. The gas washer removes a high percentage of the dust before the gases pass through the induced draft fan and enter the chimney.

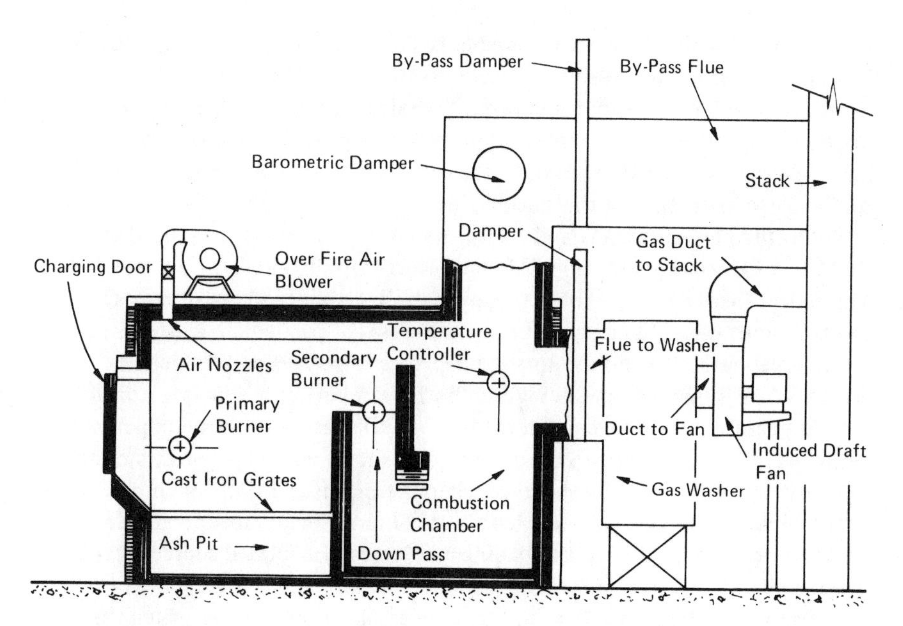

Figure 11-1. Conventional On-Site Incinerator. Source: Incinerator Institute of America, *Modern Incineration: A Solution–Not the Problem*, p. 5.

"Starved-Air" Incinerator System

While this is basically a two-chamber burning system, the fundamental departure from the conventional type is a more restrictive control of the air supply in the sealed primary chamber so as to entrain a minimum of fly ash. Burning takes place on a hearth rather than a grate. Metered air is supplied to strategic zones. Gases generated in the primary chamber are burned to completion in a high-turbulence secondary chamber, which is maintained at about 1600 degrees F. by an oil or gas burner. The result is clean stack exhaust that usually does not require further cleaning to meet or exceed pollution limits.

The advantages of a starved-air incinerator are maximum reduction of the waste by the high temperatures and minimum pollution because of the low turbulence. The starved-air chamber also conserves fuel and aids in self-sustained burning. There are certain restrictions in such a system. The chambers or furnaces need to be sealed to prevent ambient air from reaching the burning refuse. Also, after a batch is fed to the chamber, the charging door must not be opened as the induced draft will upset the equilibrium of the system. In the larger units, successive refuse charges are introduced by a ram without upsetting the air flow.

The controlled-air incinerators can be designed and operated such that there is little or no air pollution involved. This is accomplished by the fact that the burning in the primary chamber is a quiescent pyrolysis type process in which the air is controlled to partially oxidize the material into a burnable gas that is completely combusted in the turbulent secondary chamber. The important point is that the quiescent primary chamber retains the dust and ash while the gases are drawn out to the secondary chamber, avoiding the problem of particulates at the stack outlet.

A controlled-air, two-chamber, low-pollution incinerator produced by the Ross Engineering Division of Midland-Ross Corp. is shown in Figure 11-2. This unit utilizes a high-temperature, piggy-back chamber to assist in more complete burning. The principle of operation is that particulate matter drops out of an air stream when the air stream changes direction and reduces its velocity. This principle operates within both chambers of the incinerator shown.

In operation, the system is turned on and the gas flame in the upper chamber ignites. Preheating of this chamber assures that it will be able to function properly when the refuse in the lower chamber begins to burn. An air curtain at the charging door also starts. This source of air protects the operator from flames blowing back during the loading and serves as an added source of combustion air.

Overfire air enters the rear of the main combustion chamber when the upper-chamber temperature reaches 600 degrees F. Underfire air turns on automatically when the temperature of the upper chamber reaches 1500 de-

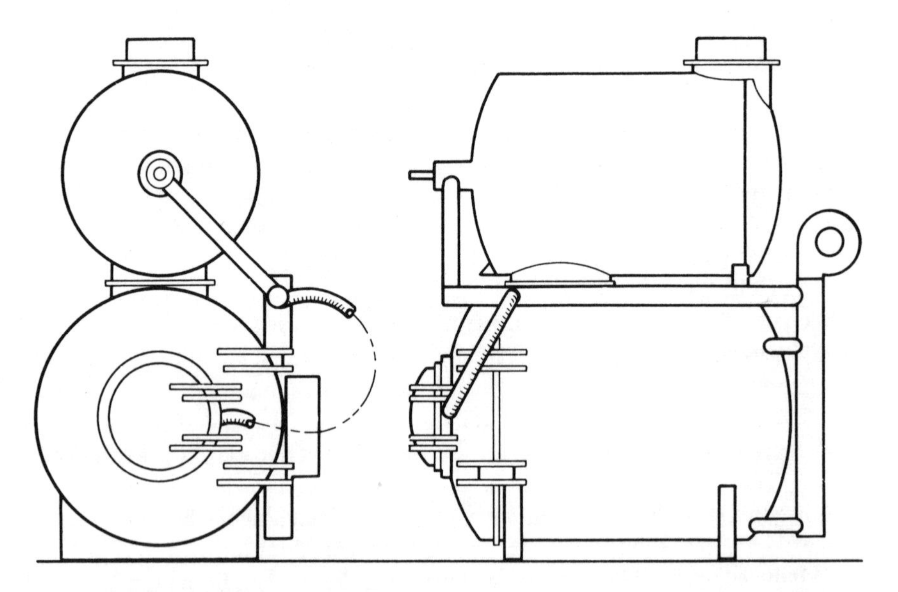

Figure 11-2. Two-Chamber Controlled-Air Incinerator. Source: Ross Engineering Division, Mid-Land Ross Corporation.

grees F. The combination of underfire and overfire air assists in the burning process and results in maximum possible incineration of the trash. Smoke rises from the lower chamber to the upper chamber where it encounters heat and added turbulence.

When the upper chamber reaches 1500 degrees F., the gas burner turns off, but the air supply to this portion continues producing turbulence along the chamber. Additional burning is promoted by this air; its angle of entry increases the turbulence. Air travel is slowed to nine feet per second, allowing a retention time of two-thirds of a second for complete consumption of smoke, odors, and gases.

When the main combustion chamber reaches 1950 degrees F., the underfire air is turned off while the air curtain and overfire air continue. This provides cooling to protect the refractory lining. A warning light cautions the operator against loading more refuse until the temperature subsides.

In many tests, the stack effluent was below that allowable by the Los Angeles APCD regulations.

Flue-Gas Scrubbers

Some older on-site incinerators can be improved by the addition of accessory equipment available from various manufacturers. One item that can be attached to some systems now incapable of meeting the more rigid air pollution standards is a flue gas scrubber. The three important considerations in attaching such a device are available space, a supply of water, and a drain capable of handling the waste water from the scrubber.

In the smaller incinerators, the water run-off is so low that it is not typically regulated by municipalities. The concentration of acid in the water is kept at a minimum by 100 percent makeup of fresh water.

In case of power failure while a charge is burning, it is advisable to have an automatic by-pass of the gases to the stack, or a vent on the secondary combustion chamber.

Fume Incinerators

Volatile fumes from industrial processes also require some form of elimination prior to exhausting into the atmosphere. Frequently such fumes may be eliminated by a reprocessing procedure where the final product has some marketable value. Otherwise, a fume incinerator of the type shown in Figure 11-3 may be used. In this device, the fumes are passed through a gas burner that operates over an adjustable range of 1200 to 1800 degrees F. The exhaust stack above the burner must be of some refractory material to withstand the

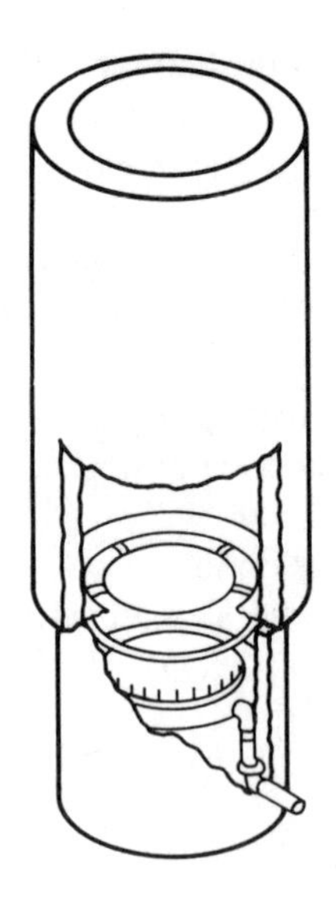

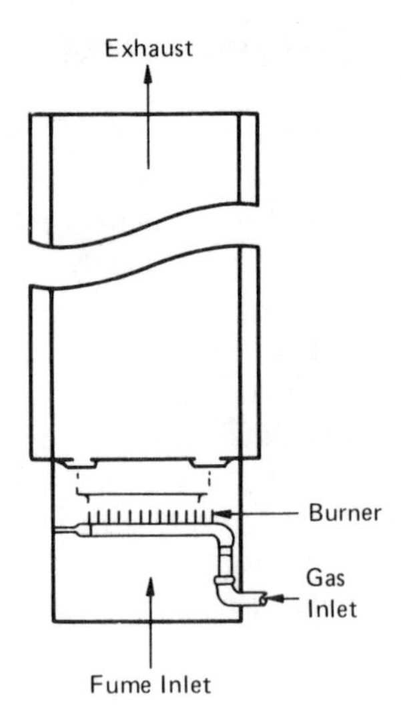

Figure 11-3. Fume Incinerator for Combustion of Fumes, Gases, and Vapors. Source: Surface Combustion Division, Mid-Land Ross Corporation.

high temperatures. Inlet exhaust gas (or fume) temperatures up to 800 degrees F. can be accommodated in this stack incinerator produced by the Surface Combustion Division of Midland-Ross Corporation.

Economics of On-Site Incineration

The capital investment per ton per day of an on-site incinerator will be related to the efficiency of the system, pollution controls required, expected life, local labor, and various other costs. The majority of the on-site incinerators sell for $1500 to $2500 per ton per day of trash handled. This price does not include installation, any automatic loading equipment, chimney, or other accessories that could be attached to the basic incinerator. There are some minimal incinerators that sell in the $800-per-ton-per-day range, but these are inferior as far as air pollution control is concerned. Some of the larger incinerators (above 2000 pounds per hour) cost more than $2500 per ton per day; these are very similar to municipal systems. (Also, it should be noted that a

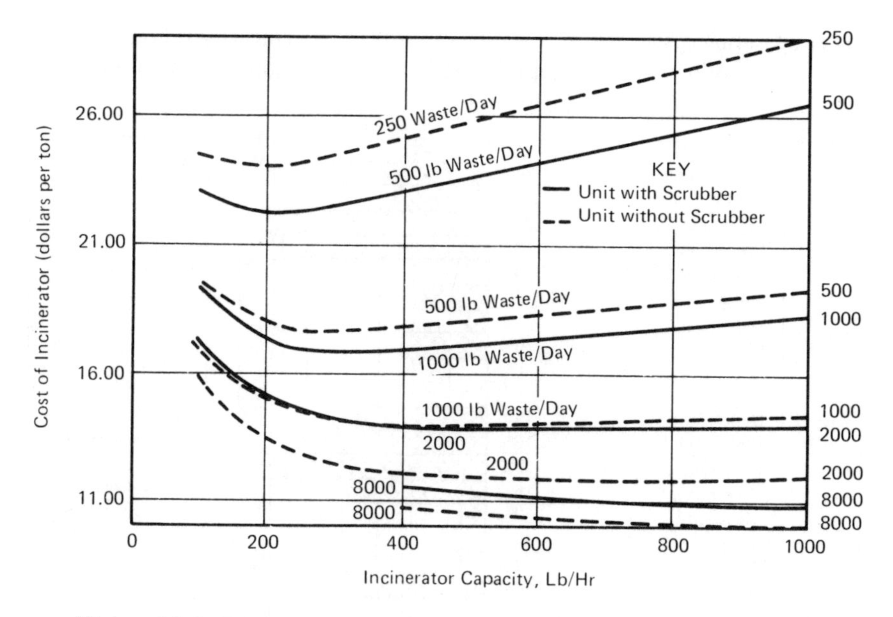

Figure 11-4. Incinerator Owning and Operating Costs. Source: R.E. Zinn and W. Niessen, "Commercial Incinerator Design Criteria," *Proceedings of the National Incinerator Conference* (1968), ASME, p. 342.

2000-pound-per-hour incinerator would easily handle all the trash from a small community of 3200 people if the unit were operated only 8 hours per day.)

In an analysis of costs of smaller incinerators, Zinn and Niessen indicated that systems capable of handling less than 200 pounds of trash per hour can be very uneconomical.[3] The cost for owning and operating incinerators of various hourly capacities is shown for different loadings in Figure 11–4. These curves indicate that the per-ton cost of incineration drops quite rapidly as the quantity of trash being handled increases. The per-ton cost of handling 500 pounds per day is twice that of handling 8000 pounds per day. A more dramatic contrast in costs is shown in the distributions of Figure 11-5, which compares a very low volume of 250 pounds of waste burned per day and a higher 2000 pounds per day.

Unfortunately, economics usually favor a larger incinerator operating fewer hours per day than a small incinerator operating continuously at peak capacity. The capital cost of the lower incinerator will be higher overall, but lower operating costs per ton, lower labor costs from a single eight-hour shift operation, and lower per-ton capital costs could make a joint venture profitable.

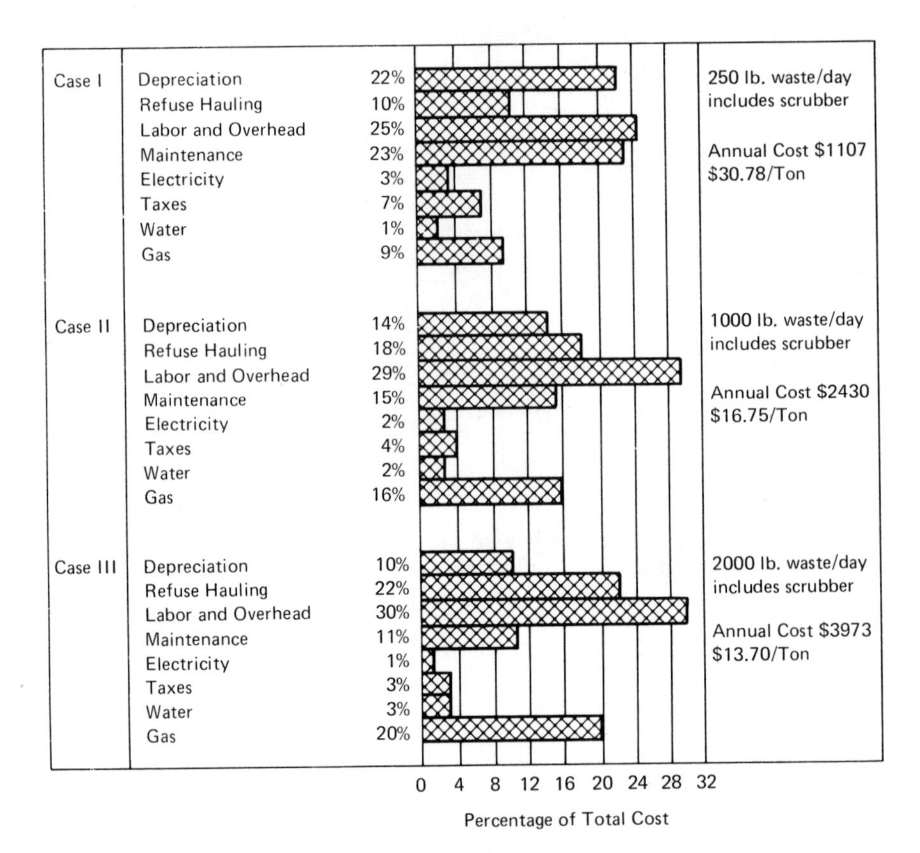

Figure 11-5. Distribution of Costs for Small Incinerators. Source: R.E. Zinn and W. Niessen, "Commercial Incinerator Design Criteria," *Proceedings National Incinerator Conference* (1968), ASME, p. 344.

Notes

Chapter 1
Introduction

1. On the other hand, "The Practice of Refuse Incineration in Japan; Burning of Refuse with High Moisture Content and Low Calorific Value," by K. Matsumoto, R. Asukata, and T. Kawashima, *Proceedings of the National Incinerator Conference (1968),* presents an overview of incineration practices in Japan, where refuse is much higher in moisture level and has about half the heat value per pound of U.S. incinerators. As a result, the Japanese incinerators require larger grate area and longer residence time.

Chapter 2
Current Incinerator Practice

1. Rodney R. Fleming, "Solid Waste Disposal, Part II—Incineration and Composting," *The American City,* Vol. 81, February 1966, pp. 94–96. Copyright © Buttenheim Publishing Company.
2. F.L. Heaney, "Regional Districts for Incineration," *National Incinerator Conference Proceedings* (1968), pp. 123–128.
3. Jack DeMarco *et al., Incinerator Guidelines—1969,* U.S. Public Health Service, Department of HEW, 1969, p. 11.

Chapter 3
Current Incinerator Design

1. C.A. Rogus, "Incinerator Design," *The American City,* Vol. 77, No. 4, April 1962, pp. 104–106. Copyright © Buttenheim Publishing Company.

Chapter 4
Refuse Selection and Preparation Before Burning

1. Grabs, the mechanical lifting hands that actually move the refuse, are thoroughly discussed in "The Operational Characteristics of Refuse Handling Grabs," by Peter J. Scott and John R. Holmes. This paper, from the *Proceedings of the National Incinerator Conference* (1972), is an overview of the state of the art. More accurately, a grab is the bucket that comes down from the overhead crane and dips into the refuse pit, scoops up the

refuse, and then carries it to the incinerator chute where it is dropped. Numerous operating curves and characteristics are presented in this paper.

Chapter 5
Salvage Values from Incinerator Residue

1. C.B. Kenehan *et al., Composition and Characteristics of Municipal Incinerator Residues,* 1968, pp. 2–3.
2. *Ibid.,* p. 19.
3. P.M. Sullivan and M.H. Stanczyk, *Economics of Recycling Metals and Minerals from Urban Refuse,* Technical Progress Report 33, U.S. Department of the Interior, Bureau of Mines, 1971.
4. National Center for Resource Recovery, *Resource Recovery from Municipal Solid Waste* (Lexington, Mass.: Lexington Books, D.C. Heath and Company, 1974).

Chapter 6
Representative Incinerator Operations

1. All plant specifications are taken from: Erwin G. Hansen and Henri Rousseau, "An Engineering Approach to the Waste Disposal Crisis," *Combustion,* March 1970, pp. 8–13.
2. N.L. Drobny, H.E. Hull and R.F. Testin, "Navy to Incinerate Rubbish for Power," *Refuse Removal Journal,* April 1967, pp. 18, 34, 67.
3. "Northwest Incinerator Plant is the Largest Complex of Kind in Western Hemisphere," *Solid WasteManagement/RRJ,* May 1971, pp. 74, 150, 152.
4. "Trash to Ashes," *Civic,* March 1972.
5. G.L. Sutin, "The East Hamilton Solid Waste Reduction Unit," *Engineering Digest,* Vol. 15, No. 7, August 11, 1969, pp. 47–51.
6. Remarks by Maurice J. Wilson, Systems Design Consultant, I.C. Thomasson and Associates, at the Park Land Hotel, New York City, July 12, 1972.
7. "Using Wastes as Fuel, City Plant Will Provide Heating and Airconditioning," *Solid Wastes Management Magazine,* September 1972, pp. 16ff.
8. Maurice J. Wilson and Adrian M. Gammill, "Putting Solid Waste to Work in Nashville," *Professional Engineer,* October 1971.
9. Wilson, *op. cit.*
10. *Ibid.*

Chapter 7
Heat Recovery

1. W.R. Niessen and S.H. Chansky, "Nature of Refuse," *Proceedings of the National Incinerator Conference* (1970).

2. Jack DeMarco *et al., Incinerator Guidelines–1969,* U.S. Public Health Service, Department of HEW, 1969, p. 38.
3. Erwin G. Hansen and Henri Rousseau, "An Engineering Approach to the Waste Disposal Crisis," *Combustion,* Vol. 41, March 1970, p. 9.
4. DeMarco, *op. cit.,* p. 38.
5. R.S. Rochford, personal communication with R.M. Vancil, dated April 25, 1973.
6. "U.S. Lags in Using Heat Content of Solid Waste in Incineration Systems," *Solid Waste Report,* June 14, 1971, p. 106.
7. See their article in the *Proceedings of the National Incinerator Conference* (1972).
8. See *Proceedings of the National Incinerator Conference* (1968), pp. 2781–2786.
9. See *Proceedings of the National Incinerator Conference* (1970).
10. W.R. Niessen and A.F. Alsobrook, "Municipal and Industrial Refuse: Compositions and Weights," *Proceedings of the National Incinerator Conference* (1972).
11. Day and Zimmermann Associates, *Special Studies for Incinerators,* U.S. Public Health Service, Department of HEW, 1968, p. 70.
12. Connecticut Department of Environmental Protection, letter from Harold E. Francis, Solid Waste Section, March 17, 1972.
13. J.W. Regan, "Generating Steam from Prepared Refuse," *Proceedings of the National Incinerator Conference* (1970), ASME, p. 216.
14. *Ibid.*

Chapter 8
Air Pollution Control

1. E.R. Kaiser and A.A. Carotti, "Municipal Incineration of Refuse with Two Percent and Four Percent Additive of Four Plastics," *Proceedings of the National Incinerator Conference* (1972), pp. 230–244.
2. Jack DeMarco *et al., Incinerator Guidelines–1969,* U.S. Public Health Service, Department of HEW, 1969.
3. W.R. Niessen and B. Sarofim, "Incinerator Air Pollution: Facts and Speculation," *Proceedings of the National Incinerator Conference* (1970), ASME, p. 180.
4. Chem Systems, Inc., *Opportunities in Solid Waste Management,* 2 Volumes; and Jack DeMarco *et al., op. cit.*
5. "Stationary Source Air Pollution Control Techniques and Practices in the United States," (Combustion Engineering, Windsor, Conn., 1970). This paper is a valuable overview of existing technology.
6. Day and Zimmermann Associates, *Special Studies for Incinerators,* U.S. Public Health Service, Department of HEW, 1968.
7. P.D. Miller *et al., Corrosion Studies in Municipal Incinerators,* Report SW-72-3-3, U.S. Environmental Protection Agency, 1972, p. 81.
8. *Ibid.*

9. K. Thoemen, "Contribution to the Control of Corrosion Problems on Incinerators with Water-Wall Steam Generators," *Proceedings of the National Incinerator Conference* (1972), pp. 310–318.
10. R.S. Rochford, personal communication with R.M. Vancil, dated April 25, 1973.
11. Eugene Backus, "Municipal Incinerator Defeats Pollution/Corrosion Problems," *Civil Engineering–ASCE,* December, 1971, pp. 48–49.

Chapter 9
General Economic Considerations

1. Eric R. Zausner, *An Accounting System for Incinerator Operations,* U.S. Public Health Service, Department of HEW, 1970.
2. *Ibid.,* p. 1.

Chapter 10
New Developments

1. Illinois Institute of Technology Research Institute–American Thermogen Inc., "Urban Ore," 1972.
2. From R. Zinn, C.R. LaMantia, and W.R. Niessen, "Total Incineration," *Proceedings of the National Incinerator Conference* (1970) ASME, p. 116. Another of the many articles on this concept is: R. Zinn and C. LaMantia, "Total Incineration," *Waste Processing,* July 1970, p. 29.
3. J.W. Regan, "Generating Steam from Prepared Refuse," *Proceedings of the National Incinerator Conference* (1970), ASME, p. 218.
4. "Steam Generation from Solid Wastes," by L.J. Cohan, presented at Connecticut Clean Power Symposium, Combustion Engineering Corporation, 1972.
5. R.C. Bailie, "Solid Waste Incineration in Fluidized Beds," *Industrial Water Engineering,* November 1970, pp. 22–25.
6. "Pyrolysis of Refuse Gains Ground," *Environmental Science and Technology,* Vol. 5, April 1971, pp. 310–312.

Chapter 11
On-Site Incineration

1. Incinerator Institute of America, 1 Stone Place, Bronxville, New York 10708.

2. "Modern Incineration, A Solution–Not the Problem," *Bulletin of the Incinerator Institute of America,* 1 Stone Place, Bronxville, New York, 1969.
3. R.E. Zinn and W. Niessen, "Commercial Incinerator Design Criteria," *Proceedings of the National Incinerator Conference* (1968), pp. 337–354.

Bibliography

"A Zero-Residue Incinerator?" *The American City,* April 1971.

American Public Works Association. *Municipal Refuse Disposal.* Second Edition. Chicago: Public Administration Service, 1966.

American Society of Mechanical Engineers. *Proceedings of the National Incinerator Conference.* New York, 1964.

——. *Proceedings of the National Incinerator Conference.* New York, 1966.

——. *Proceedings of the National Incinerator Conference.* New York, 1968.

——. *Proceedings of the National Incinerator Conference.* New York, 1970.

——. *Proceedings of the National Incinerator Conference.* New York, 1972.

"Amsterdam's Automated Incinerator." *Public Works,* August 1970.

"An Engineering Approach to the Waste Disposal Crisis." *Secondary Raw Materials,* Vol. 6, No. 11, November 1968.

"An Engineering Approach to the Waste Disposal Crisis." Part 2. *Secondary Raw Materials,* Vol. 6, No. 12, December 1968.

"Are These Incinerators the Answer to Plastics Waste?" *Modern Plastics,* October 1970, p. 102.

"At Last–Truly Smokeless Incineration." *Modern Materials Handling,* Vol. 24, December 1969, p. 60.

Babcock & Wilcox. *Interface,* No. 11, September–October 1972.

Backus, Eugene. "Municipal Incinerator Defeats Pollution/Corrosion Problems." *Civil Engineering–ASCE,* December 1971, pp. 48–49.

Bailie, R.C. "Solid Waste Incineration in Fluidized Beds." *Industrial Water Engineering,* November 1970, pp. 22–25.

Bender, R.J. "Are European Incinerators Setting Pace?" *Power,* February 1970, p. 40.

Bremser, L.W. "Incineration." *Proceedings of the National Conference on Solid Waste Research,* p. 108. U.S. Public Health Service and American Public Works Association. Chicago, 1963.

Browning, J.E. "Fluidized-Bed Combustion Broadens Its Horizons." *Chemical Engineering,* June 28, 1971, pp. 44–46.

Burns. "Precision Controlled Flames Fight Air Pollution." *The American City, The American City,* Vol. 85, No. 5, May 1970, pp. 98–102.

Chem Systems, Inc. *Opportunities in Solid Waste Management.* 2 Volumes.

"Chicago and London Take the Financial Sting Out of Garbage." *Engineering News-Record,* Vol. 183, No. 23, December 4, 1969, pp. 38–39.

Cohan, L.J. "Steam Generation From Solid Wastes." Presented at Connecticut Clean Power Symposium. Combustion Engineering Corp., 1972.

Connecticut Department of Environmental Protection. Letter from Harold E. Francis, Solid Waste Section, March 17, 1972.

Copeland, G.G. "The Design and Operation of Fluidized Bed Incinerators for Solid and Liquid Wastes." Presented at the National Industrial Solid Waste Management Conference, University of Houston, March 1970.

Corey, Richard C. *Principles and Practices of Incineration.* New York: Wiley-Interscience, 1969.

"Corps Hearing Sought on Proposed Hempstead Harbor Disposal of Incinerator Ash." *Solid Waste Report,* January 25, 1971, p. 16.

Crawford, Gary N. "Solid Waste Incinerators–Applications and Limitations." *Plant Engineering,* March 19, 1970, pp. 62–63.

——."Solid Waste Incinerators–Design and Operation." *Plant Engineering,* April 2, 1970, pp. 35–37.

Cross, Frank L., Jr. *Handbook On Incineration: A Guide to Theory, Design, and Maintenance.* Westport, Connecticut: Technomic Publishing Co., Inc., 1972.

Day and Zimmermann Associates. *Special Studies for Incinerators.* U.S. Public Health Service, Department of HEW. 1968.

DeMarco, Jack, *et al. Incinerator Guidelines–1969.* U.S. Public Health Service, Department of HEW. 1969.

DeVon Bogue, M. "Municipal Incineration." Presented at the New York State Health Department In-Service Training Course "Refuse Disposal by Sanitary Landfill and Incineration," December 1, 1965, Albany, N.Y., U.S. Public Health Service.

"Dow Receives Patent for Process Controlling Pollution from Plastic Incineration." *Solid Waste Report,* January 25, 1971, p. 13.

Drobny, N.L., Hull, H.E., and Testin, R.F. "Navy to Incinerate Rubbish for Power." *Refuse Removal Journal,* April 1967.

"Dual Chamber Incinerator Burns Up Problem Wastes." *Chemical Engineering,* April 6, 1970, p. 68.

Duncan, L.P. *Analysis of Final State Implementation Plan–Rules and Regulations.* EPA Report APTD-1334. July 1972.

Edwards, Geoff. "Refuse Disposal by Burning." *Engineering,* September 12, 1969.

"Electricity From the Dustbin." *Nature,* Vol. 222, May 31, 1969, pp. 812–813.

"Electricity From the Rubbish Heap." *Electronics and Power,* Vol. 16, September 1970, p. 345.

Engdahl, Richard B. *Solid Waste Processing–A State-of-the-Art Report on Unit Operation and Processes.* U.S. Public Health Service, Department of HEW. 1969.

"Estimates of Solid Waste Management Costs." *Solid Waste Report,* June 14, 1971, p. 108.

Ewing, Robert C. "Smokeless Fluid Bed Incinerator." *The Oil and Gas Journal,* December 15, 1969, pp. 68–69.

Faaty, A.C., Jr. "Centralized Waste Disposal Facility is Economical." *The Oil and Gas Journal,* August 11, 1969, pp. 142–144.

Fernandes, J.H. "Stationary Source Air Pollution Control Techniques and Practices in the United States." Presented at the Industrial Air Treatment and Pollution Control Equipment Symposium, Frankfurt, Germany, November 10–12, 1970. Available from Combustion Engineering, Inc.

Fleming, Rodney R. "Solid Waste Disposal, Part II–Incineration and Composting." *The American City,* Vol. 81, February 1966, pp. 94–96.

"Futuristic Incinerator Planned for Hamilton." *Solid Wastes Management Magazine,* Vol. 12, March 1969, pp. 14, ff.

Garrett, C.J. "Incineration Control." *Industrial Water Engineering,* July 1970, pp. 35–39.

"Giant Eidal Disposal Machine Sold to Eastman Kodak Company." *Scrap Age,* July 1969.

"Giant Incinerator Solves Problem." *The Engineer,* July 3, 1969, p. 9.

Hansen, Erwin G., and Rousseau, Henri. "An Engineering Approach to the Waste Disposal Crisis." *Combustion,* Vol. 41, March 1970, pp. 8–13.

Henn, John J. and Peters, Frank A. *Cost Evaluation of a Metal and Mineral Recovery Process for Treating Municipal Incinerator Residues.* Information Circular 8533. U.S. Department of the Interior, Bureau of Mines, 1971.

"Hico Corp. Gets $5 Million Order to Replace NYC Incinerators with Compactors." *Solid Waste Report,* April 19, 1971, p. 72.

"High Temperature Garbage Disposal Turns Up Iron, Glass." *Industry Week,* Vol. 166, No. 9, March 2, 1970.

Hoffman, Donald A. and Fitz, Richard A. "Batch Retort Pyrolysis of Solid Municipal Wastes." *Environmental Science & Technology,* Vol. 2, No. 11, November 1968, pp. 1023–1026.

Illinois Institute of Technology Research Institute—American Thermogen, Inc. "Urban Ore." 1972.

"Incinerator Emits No Visible Smoke." *Building Systems Design,* Vol. 67, July 1970, pp. 20–21.

Incinerator Institute of America. *I.I.A. Incinerator Standards.* New York, N.Y., November, 1968.

"Incinerator Reduces Waste Disposal Cost." *Plant Engineering,* January 8, 1970, p. 18.

"Incinerators, Environment, and Other Radical Subjects." *The American City,* Vol. 85, No. 5, May 1970, p. 8.

"Install Precipitation Units in NYC's Incinerators." *Secondary Raw Materials,* August 1968, p. 35.

Kaiser, E.R. and McCaffery, Joseph B. "Overfire Air Jets for Incinerator Smoke Control." *Combustion,* Vol. 42, August 1970, pp. 20–22.

"Keep Britain Tidy." *Solid Wastes Management Magazine,* Vol. 13, No. 1, January 1970, p. 14.

Keith, Joseph A. "Pawtucket's Two-Pronged Attack." *The American City,* August 1966, pp. 108–109.

Kenahan, C.B. *et al. Composition and Characteristics of Municipal Incinerator Residues.* Report 7204. U.S. Department of the Interior, Bureau of Mines, 1968.

Kramer, W.P. "The Capabilities and Limitations of Current Incinerator Designs." *Public Works,* Vol. 101, No. 7, July 1970, pp. 57–59.

LaRue, Phillip G. "Pollution-Controlled Gas Incineration." *ASHRAE Journal,* February 1970, pp. 58–62.

Lucier, T.E. "The Pit Incinerator." *Industrial Water Engineering,* September 1970, pp. 28–30.

Mallatt, R.C. *et al.* "Incinerate Sludge and Caustic." *Hydrocarbon Processing,* May 1970, pp. 121–122.

"Melting Waste Into 'Urban Ore'." *Business Week,* July 25, 1970, p. 98.

Michaels, Abraham. "Only 15% of All Solid Waste Is Incinerated." *Refuse Removal Journal,* Vol. 10, No. 2, February 1967, pp. 20–22.

Miller, P.D. *et al. Corrosion Studies in Municipal Incinerators.* Report SW-72-3-3. National Environmental Research Center, U.S. Environmental Protection Agency, Cincinnati, Ohio. 1972.

"Modern Incineration, A Solution—Not the Problem." *Bulletin of Incinerator Institute of America.* Bronxville, New York, 1969.

Muhich, A.J. *et al. Preliminary Data Analysis, 1968 National Survey of Community Solid Waste Practices.* U.S. Public Health Service, Department of HEW. 1968.

National Center for Resource Recovery. *Resource Recovery from Municipal Solid Waste.* Lexington, Mass.: Lexington Books, D.C. Heath and Company, 1974.

"New Concept in Incineration." *The American City,* Vol. 85, No. 5, May 1970, p. 62.

"New Incentives Needed for Disposing of Hospital Solid Wastes." *Solid Waste Report,* May 3, 1971, pp. 77–78.

"New Incinerators Meet Anti-Pollution Laws." *Material Handling and Engineering,* September 1969, pp. 107–108.

"New Twist in Waste Use." *Chemical Week,* Vol. 106, No. 2, January 14, 1970, p. 58.

Niessen, W.R. "Incinerator Emission Control." *Industrial Water Engineering,* August 1970, pp. 26–31.

Niessen, W.R. *et al.Systems Study of Air Pollution from Municipal Incineration.* Arthur D. Little, Inc., for the U.S. Department of HEW. March 1970.

"Northwest Incinerator Plant is Largest Complex of Kind in Western Hemisphere." *Solid Wastes Management Magazine,* May 1971, pp. 74, ff.

"Oregon's First Clean Commercial Incinerator to Start Operating Next Month." *Solid Waste Report,* January 11, 1971, p. 5.

"Plastic Incineration Problem Seen Growing With Its Percentage of Waste." *Solid Waste Report,* June 14, 1971, p. 107.

"Plastic Wastes Yield to Pyrolysis." *Environmental Science & Technology,* Vol. 4, June 1970, p. 473.

"Plastics Pollution Minimal, But—." *Industry Week,* May 11, 1970.

"Plastics Solid Waste Can Increase Incineration Efficiency, Goodrich Experts Report." *Solid Waste Report,* May 3, 1971, p. 76.

"Promising Future for Underground Incineration." *The American City,* Vol. 85, No. 5, May 1970, p. 48.

"Pyrolysis of Refuse Gains Ground." *Environmental Science & Technology,* Vol. 5, April 1971, pp. 310–312.

"Recipe for 'Cooking' Trash." *Chemical Week,* April 8, 1970, pp. 47–49.

"Record Incinerator to Burn 6,000 Tons per day." *Engineering News-Record,* August 13, 1970, p. 15.

Resource Recovery Act of 1969. Hearings on S. 2005 Before the Subcommittee on Air and Water Pollution of the Committee of Public Works, U.S. Senate. Serial No. 91-13, Washington, D.C. 1969.

Rogus, C.A. "Incinerator Design." *The American City,* Vol. 77, No. 4, April 1962, pp. 104–106.

Rohr, Fred W. "One Way to Control It–Burn It." *Actual Specifying Engineer,* Vol. 18, No. 5, November 1969, pp. 74–79.

"Sears Opposes Use of Incinerators." *Chain Store Age,* November 1970, p. E–27.

"Solid Waste Disposal System." *Pollution Equipment News,* December 1970.

Steinko, Franklin A., Jr. "The Heat's Off: Firing Date Set." *American City Government,* Vol. 34, February 1969, pp. 14–17.

Sullivan, P.M. and Stanczyk, M.H. *Economics of Recycling Metals and Minerals from Urban Refuse.* Technical Progress Report 33. U.S. Department of the Interior, Bureau of Mines, 1971.

Sutin, G.L. "The East Hamilton Solid Waste Reduction Unit." *Engineering Digest,* Vol. 15, No. 7, August 11, 1969, pp. 47–51.

Thomaides, Lazarus. "Why Catalytic Incineration?" *Pollution Engineering,* May–June 1971, pp. 32–33.

" 'Total Destruct' Waste Disposal." *Mechanical Engineering,* June 1970, p. 56.

"Trash to Ashes." *Civic,* March 1972.

Trauernicht, J. O'Rinda. "Plastics Waste: Burn It or Bury it?" *Plastics Technology,* Vol. 16, July 1970, pp. 29–31.

"Trends and Practices in Municipal Incinerator Design." *Public Works,* June 1966.

"Two New Waste Incineration Projects Announced." *Environmental Science & Technology,* Vol. 4, No. 3, March 1970, p. 184.

"U.S. Lags in Using Heat Content of Solid Waste in Incineration Systems." *Solid Waste Report,* June 14, 1971, p. 106.

"Using Wastes as Fuel, City Plant will Provide Heating and Airconditioning." *Solid Wastes Management Magazine,* September 1972, pp. 16 ff.

Wallace, Lee. "Electric Incinerators–Newcomer to the Total Electric Family." *ASHRAE Journal,* May 1970, pp. 48–49.

"Wheelabrator Awarded $1.4 Million Air Pollution System for Washington Incinerator." *Solid Waste Report,* January 11, 1971, p. 5.

Wilson, David Gordon (Ed.). *The Treatment and Management of Urban Solid Waste.* Westport, Connecticut: Technomic, 1972.

Wilson, Maurice J. and Gammill, Adrian M. "Putting Solid Waste to Work in Nashville." *Professional Engineer,* October 1971.

Witt, Philip A. "Disposal of Solid Wastes." *Chemical Engineering,* October 4, 1971.

Zausner, Eric R. *An Accounting System for Incinerator Operations.* U.S. Public Health Service, Department of HEW. 1970.

Zinn, R. and LaMantia, C. "Total Incineration." *Waste Processing,* July 1970, pp. 29, ff.

"$1.7 Million More Needed to Control Air Pollution from New D.C. Incinerator." *Solid Waste Report,* February 8, 1971.

Index

92
80
172

Low cost = [illegible]
Water & Waste = [illegible]
Water Quality = [illegible]
Chemistry = [illegible]
Advanced = [illegible]
Microbiology = [illegible]
Water Resources = [illegible]
Sed. Trans = [illegible]
Vege = [illegible]
Solid Waste = [illegible]
Hydrol. = [illegible]
Sim = [illegible]

Average = [illegible] [illegible] [illegible]
say [illegible]

∴ say [illegible] average for the last 16 subjects